SpringerWienNewYork

Dmitrij Frishman

Editor

Structural Bioinformatics
of Membrane Proteins

SpringerWienNewYork

Prof. Dmitrij Frishman
TU München, Wissenschaftszentrum Weihenstephan, Freising, Germany

© 2010 Springer-Verlag/Wien
Printed in Austria

SpringerWienNewYork is part of
Springer Science + Business Media
springer.at

Typesetting: Thomson Press (India) Ltd., Chennai
Printing: Holzhausen Druck GmbH, 1140 Wien, Austria

Printed on acid-free and chlorine-free bleached paper
SPIN: 127 63 863

With 63 (partly coloured) Figures

Library of Congress Control Number: 2010929320

ISBN 978-3-7091-0044-8 SpringerWienNewYork

CONTENTS

Topology prediction of membrane proteins: how distantly related homologs come into play (*Rita Casadio, Pier Luigi Martelli, Lisa Bartoli, Piero Fariselli*)   61

Transmembrane beta-barrel protein structure prediction (*Arlo Randall, Pierre Baldi*)   83

Multiple alignment of transmembrane protein sequences (*Walter Pirovano, Sanne Abeln, K. Anton Feenstra, Jaap Heringa*)   103

Prediction of re-entrant regions and other structural features beyond traditional topology models (*Erik Granseth*)   123

Dual-topology: one sequence, two topologies (*Erik Granseth*)   137

Predicting the burial/exposure status of transmembrane residues in helical membrane proteins (*Volkhard Helms, Sikander Hayat, Jennifer Metzger*)   151

Helix–helix interaction patterns in membrane proteins (*Dieter Langosch, Jana R. Herrmann, Stephanie Unterreitmeier, Angelika Fuchs*)   165

Predicting residue and helix contacts in membrane proteins (*Angelika Fuchs, Andreas Kirschner, Dmitrij Frishman*)   187

Prediction of three-dimensional transmembrane helical protein structures (*Patrick Barth*)   231

GPCRs: past, present, and future (*Bas Vroling, Robert P. Bywater, Laerte Oliveira, Gert Vriend*)   251

Evolutionary origins of membrane proteins

Armen Y. Mulkidjanian[1,2] **and Michael Y. Galperin**[3]

[1]School of Physics, University of Osnabrück, Osnabrück, Germany
[2]A.N. Belozersky Institute of Physico-Chemical Biology, Moscow State University, Moscow, Russia
[3]National Center for Biotechnology Information, National Library of Medicine, National Institutes of Health, Bethesda, MD, USA

Abstract

Although the genes that encode membrane proteins make about 30% of the sequenced genomes, the evolution of membrane proteins and their origins are still poorly understood. Here we address this topic by taking a closer look at those membrane proteins the ancestors of which were present in the Last Universal Common Ancestor, and in particular, the F/V-type rotating ATPases. Reconstruction of their evolutionary history provides hints for understanding not only the origin of membrane proteins, but also of membranes themselves. We argue that the evolution of biological membranes could occur as a process of co-evolution of lipid bilayers and membrane proteins, where the increase in the ion-tightness of the membrane bilayer may have been accompanied by a transition from amphiphilic, pore-forming membrane proteins to highly hydrophobic integral membrane complexes.

1 Introduction

The origins of membrane proteins are inextricably coupled with the origin of lipid membranes. Indeed, membrane proteins, which contain hydrophobic stretches and are generally insoluble in water, could not have evolved in the absence of functional membranes, while purely lipid membranes would be impenetrable and hence useless without membrane proteins. The origins of biological membranes – as complex cellular devices that control the energetics of the cell and its interactions with the surrounding world (Gennis 1989) – remain obscure (Deamer 1997; Pereto et al. 2004).

Corresponding author: Armen Y. Mulkidjanian, School of Physics, University of Osnabrück, 49069 Osnabrück, Germany (E-mail: amulkid@uos.de)

1

The traditional approach that is employed to reconstruct the early evolution of a particular cellular system is to compare the complements of its components in bacteria and archaea, the two domains of prokaryotic life (Koonin 2003). The conservation of a set of essential genes between archaea and bacteria leaves no reasonable doubt in the existence of some version of Last Universal Common Ancestor (LUCA) of all cellular organisms (Koonin 2003; Glansdorff et al. 2008; Mushegian 2008). The comparison of particular cellular systems in bacteria and archaea yielded informative results, especially, in the case of the translation and the core transcription systems (Harris et al. 2003; Koonin 2003). However, the comparison of bacteria and archaea does not shed light on the origin of biological membranes because they fundamentally differ in these two domains of prokaryotic life (Wächtershäuser 2003; Boucher et al. 2004; Pereto et al. 2004; Koonin and Martin 2005; Koga and Morii 2007; Thomas and Rana 2007). The dichotomy of the membranes led to the proposal that the LUCA lacked a membrane organization (Martin and Russell 2003; Koonin and Martin 2005). However, the nearly universal conservation of the key subunits of complex membrane-anchored molecular machines, such as general protein secretory pathway (Sec) system (Cao and Saier 2003) and the F/V-type ATP synthase (Gogarten et al. 1989; Nelson 1989), indicates that LUCA did possess some kind of membrane (Koonin and Martin 2005; Jekely 2006).

The universal conservation of the key subunits of the F/V-type ATPases/synthases (F/V-ATPases) – elaborate, rotating molecular machines that couple transmembrane ion transfer with the synthesis or hydrolysis of ATP (see Boyer 1997; Walker 1998; Perzov et al. 2001; Müller and Gruber 2003; Weber and Senior 2003; Yokoyama and Imamura 2005; Beyenbach and Wieczorek 2006; Dimroth et al. 2006; Forgac 2007; Mulkidjanian et al. 2009, for reviews) – is particularly challenging, since this enzyme complex is apparently built of several modules (Walker 1998) and therefore is anything but primitive. Therefore F/V-type ATPases, together with the related bacterial flagella, make one of the main exhibits of today's proponents of "Intelligent Design". The F-type and V-type ATPases are also remarkable as being one of the few cases, outside the translation and core transcription systems, where the classic, "Woesian" phylogeny (Woese 1987) is clearly seen, with the primary split separating bacteria from the archaeo-eukaryotic branch that splits next (Gogarten et al. 1989; Nelson 1989). F/V-type ATPases are more "demanding" than the Sec system – they require perfect, ion-tight membranes for proper functioning. Hence understanding the evolution of the F/V-type ATPases might shed light on the evolution of not only membrane proteins but also membranes proper.

Recently, by combining structural and bioinformatics analyses, we addressed the evolution of the F/V-type ATPases by comparing the structures and sequences of archaeal and bacterial members of this class of enzyme (Mulkidjanian et al. 2007, 2008a,b, 2009). Here we survey these findings and explore their implications for the origin and the earliest evolution of membranes. We argue that the history of membrane enzymes was essentially shaped by the evolution of membranes themselves. In addition, we discuss the mechanisms of an evolutionary transition between the primitive replicating entities and the first membrane-encased life forms, as well as the role of mineral compartments of hydrothermal origin in this transition.

2 Comparative analysis of F/V-type ATPases: example of function cooption?

Together with two evolutionarily unrelated families, the P-type ATPases and ABC transporters, the F/V-type ATPases belong to a heterogeneous group of enzymes that use the energy of ATP hydrolysis to translocate ions across membranes (Skulachev 1988; Gennis 1989; Cramer and Knaff 1990; Saier 2000). The F/V-type ATPases, however, are unique functionally, because they can efficiently operate as ATP synthases, and mechanistically, in that their reaction cycle is accompanied by rotation of one enzyme part relative to the other (Noji et al. 1997; Imamura et al. 2005). Biochemically, F/V-ATPases are composed of membrane-bound parts (F_O and V_O, respectively) and catalytic protruding segments (F_1 and V_1), which can be washed off the membrane, e.g., by Mg^{2+}-free solution (see Fig. 1a). The headpiece of the better studied F-type ATPases is a hexamer of three α- and three β-subunits with each of the latter carrying an ATP/ADP-binding catalytic site (Stock et al. 2000). The hexamer, together with the peripheral stalk and the membrane anchor, makes the "stator" of this enzyme complex. The "rotor" consists of the elongated γ-subunit that, via the globular ε-subunit, is connected to a ring-like oligomer of 10–15 small c-subunits (see Fig. 1 and Deckers-Hebestreit et al. 2000; Gibbons et al. 2000; Stock et al. 2000; Capaldi and Aggeler 2002; Angevine et al. 2003; Pogoryelov et al. 2005). The sequential hydrolysis of ATP molecules by the $\alpha_3\beta_3$ catalytic hexamer rotates the central stalk together with the ring of c-subunits relative to the stator, so that the ring slides along the membrane subunits of the stator (Boyer 1997; Noji et al. 1997; Panke et al. 2000; Itoh et al. 2004). This sliding movement is coupled to the transmembrane ion transfer and generation of membrane potential (Cherepanov et al. 1999; Mulkidjanian 2006). The enzyme also functions in the opposite direction, i.e., as an ATP synthase. In this mode, the ion current rotates the c-ring, and the ATP synthesis is mediated by sequential interaction of the rotating γ-subunit with the three catalytic β-subunits (Cherepanov et al. 1999; Capaldi and Aggeler 2002; Weber and Senior 2003). The V-type ATPases share a common overall scaffold with the F-ATPases but differ from them in many structural and functional features (for

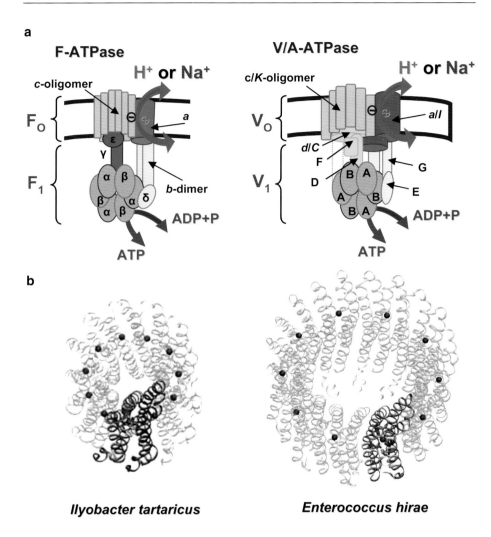

Fig. 1.   Structure and evolutionary relationships of F-type and A/V-type ATPases. (**a**) Modern F-type and V-type ATPases; the minimal, prokaryotic sets of subunits are depicted; orthologous subunits are shown by the same colors and shapes, and non-homologous but functionally analogous subunits of the central stalk are shown by different colors. The a-subunits that show structural similarity but might not be homologous (Mulkidjanian et al. 2007) are shown by distinct but similar colors; in the case of those V-ATPase subunits that are differently denoted in prokaryotes and eukaryotes, double notation is used: eukaryotic/prokaryotic. The composition of peripheral stalk(s) and their number in V-ATPases remains ambiguous, with values of up to 3 being reported (Esteban et al. 2008; Kitagawa et al. 2008). For further details, see refs. (Mulkidjanian et al. 2007, 2009). (**b**) Membrane rotor subunits of the Na$^+$-translocating ATP synthases; left, undecamer of c-subunits of the Na$^+$-translocating F-type ATP synthase of *Ilyobacter tartaricus* (PDB entry 1YCE; Meier et al. 2005); right, decamer of K subunits of the Na$^+$-translocating V-type ATP synthase of *Enterococcus hirae* (PDB entry 2BL2; Murata et al. 2005); both rings are tilted to expose the internal pore; in *I. tartaricus*, Na$^+$ ions (purple) crosslink the neighboring subunits, whereas in *E. hirea* the Na$^+$ ions are bound by four-helical bundles that evolved via a subunit duplication (see also Mulkidjanian et al. 2008b, 2009).

details see Fig. 1a and Müller and Gruber 2003; Imamura et al. 2005; Yokoyama and Imamura 2005; Drory and Nelson 2006; Mulkidjanian et al. 2007; Mulkidjanian et al. 2009).

F-type ATPases are found in bacteria and in eukaryotic mitochondria and chloroplasts, whereas the V-type ATPases are found in archaea, some bacteria, and in membranes of eukaryotic cells (Gogarten et al. 1989; Perzov et al. 2001; Nakanishi-Matsui and Futai 2006; Mulkidjanian et al. 2008b). In particular, vacuoles contain V-type ATPases that use the energy of ATP hydrolysis to acidify cellular compartments (Nelson 1989; Perzov et al. 2001; Beyenbach and Wieczorek 2006; Forgac 2007). Some authors classify the simpler, prokaryotic V-type ATPases into a separate subgroup of A-type (from archaeal) ATPases/ATP synthases (Hilario and Gogarten 1998; Müller and Gruber 2003). Others, however, prefer to speak about bacterial and eukaryotic V-type ATPases (Perzov et al. 2001; Drory and Nelson 2006; Nakanishi-Matsui and Futai 2006). In phylogenetic trees, the A-type ATPases invariably cluster together with the eukaryotic V-ATPases and separately from the F-type ATPases (Gogarten et al. 1989; Hilario and Gogarten 1993, 1998).

Among the F-type ATPases and the V-type ATPases, both proton translocating and Na^+-translocating forms are found. The ion specificity of the sodium-dependent F/V-type ATPases is, in fact, limited to the ion-binding sites of their membrane-embedded parts F_O and V_O, respectively (see Fig. 1 and von Ballmoos et al. 2008). In the absence of sodium, Na^+-ATPases have the capacity to translocate protons (Dimroth 1997; von Ballmoos and Dimroth 2007). In contrast, H^+-ATPases are apparently incapable of translocating Na^+ ions (Zhang and Fillingame 1995). This asymmetry is most likely due to the higher coordination number of Na^+, which requires six ligands (Frausto da Silva and Williams 1991), while proton, in principle, can be translocated by a single ionizable group. Comparative analyses of the subunits c of Na^+-translocating and H^+-translocating ATPases identified several residues that are involved in Na^+-binding and are the principal determinants of the coupling ion specificity (Zhang and Fillingame 1995; Rahlfs and Müller 1997; Dzioba et al. 2003). However, the exact modes of Na^+ binding in F- and V-ATPases remained obscure until the structures of the membrane-spanning, rotating c-oligomers of the Na^+-translocating ATP synthases of the F-type and V-type have been resolved (see Fig. 1b and Meier et al. 2005, 2009; Murata et al. 2005). Strikingly, the superposition of these structures reveals nearly identical sets of amino acids involved in Na^+ binding which almost perfectly superimpose in space (Mulkidjanian et al. 2008b). When pitted against the topology of the phylogenetic tree of F/V-type ATPases, the similarity of the Na^+-binding sites in the two prokaryotic domains led to the conclusion that the last common ancestor of the extant F-type and V-type ATPase, most likely, possessed a

Na$^+$-binding site (Mulkidjanian et al. 2008b). Indeed, sodium-dependent ATPases are scattered among proton-dependent ATPases in both the F-branches and the V-branches of the phylogenetic tree (Mulkidjanian et al. 2008b). Barring the extremely unlikely convergent emergence of the same set of Na$^+$ ligands in several lineages, these findings suggest that the common ancestor of F-type and V-type ATPases contained a Na$^+$-binding site.

The ion specificity of the F/V-type ATPases, however, is decisive for the nature of the bioenergetic cycle in any organism. Although proton-motive force (PMF) and/or sodium-motive force (SMF) can be generated by a plethora of primary sodium or proton pumps, F/V-type ATPases are unique in their ability to utilize PMF and/or SMF to produce ATP (Cramer and Knaff 1990). Owing to its nearly ubiquitous presence, the proton-based energetics has been generally viewed as the primary form of biological energy transduction (Deamer 1997; von Ballmoos and Dimroth 2007). By contrast, the ability of some prokaryotes to utilize sodium gradient for ATP synthesis has been usually construed as a later adaptation to survival in extreme environments (Konings 2006; von Ballmoos and Dimroth 2007). The results of our analysis indicated that the sodium-based mechanisms of energy conversion preceded the proton-based bioenergetics.

However unexpected it might be (see Skulachev 1988; Dibrov 1991; Häse et al. 2001), the evolutionary primacy of sodium bioenergetics seems to find independent support in membrane biochemistry. As argued in more detail elsewhere (Mulkidjanian et al. 2008b, 2009), creating a non-leaky membrane that can maintain a PMF sufficient to drive ATP synthesis is a harder task than making a sodium-tight membrane. The conductivity of lipid bilayers for protons is by 6–9 orders of magnitude higher than the conductivity for Na$^+$ and other small cations (Deamer 1987; Haines 2001; Konings 2006). This difference is based on the unique mechanism of transmembrane proton translocation: whereas the conductivity for other cations depends on how fast they can cross the membrane/water interface (Deamer 1987; Nagle 1987; Tepper and Voth 2006), the rate of proton transfer across the membrane is limited not by the proton transfer across the interface, but by the "hopping" of protons across the highly hydrophobic midplane of the lipid bilayer (Deamer 1987; Haines 2001). Hence, proton leakage can be suppressed by decreasing the lipid mobility in the midplane of the bilayer and/or increasing the hydrocarbon density in this region. Accordingly, proton tightness can be achieved, for example, by branching the ends of the lipid tails and/or incorporating hydrocarbons with a selective affinity to the cleavage plane of the bilayer (Haines 2001).

In agreement with the hypothesis on independent emergence of proton-based energetics in different lineages, representatives of the three domains of life employ distinct solutions to make their membranes tighter to protons, namely, the mobility of side chains is restricted in distinct ways and different hydrocarbons are packed in the midplane of the H^+-tight membranes (see Haines 2001; Konings et al. 2002; Konings 2006; Mulkidjanian et al. 2008b, for details). This fact supports the suggestion on the independent transition from the sodium to proton bioenergetics in different lineages.

Where did the first, apparently, sodium-translocating F/V-type ATPases come from? The comparison of the F-type and V-type ATPases shows that they are built of both homologous and unrelated subunits (see Fig. 1 and Mulkidjanian et al. 2007). The subunits of the catalytic hexamer and the membrane c-ring are highly conserved (Gogarten et al. 1989, 1992; Nelson 1989; Lapierre et al. 2006). The subunits that are thought to form the hydrophilic parts of the peripheral stalk(s), also appear to be homologous, despite low sequence similarity (Supekova et al. 1995; Pallen et al. 2006). The membrane parts of the peripheral stalks show structural and functional similarity as well (Kawasaki-Nishi et al. 2001; Kawano et al. 2002), although it remains unclear whether or not they are homologous. By contrast, the subunits of the rotating central shafts, which couple the catalytic hexamers with the c-ring (shown by dissimilar colors in Fig. 1), are not homologous (Nelson 1989) as substantiated by the presence of dissimilar structural folds (Mulkidjanian et al. 2007).

Building on this conservation pattern, we suggested that the common ancestor of the F-type and V-type ATP was not an ion-translocating ATPase but rather an ATP-dependent protein translocase in which the translocated protein itself occupied the place of the central stalk (Mulkidjanian et al. 2007). Indeed, the catalytic hexamers of F-type and V-type ATPases are homologous to hexameric helicases, specifically, the bacterial RNA helicase Rho, a transcription termination factor (Patel and Picha 2000). This relationship led to the earlier hypothesis that the ancestral membrane ATPase evolved as a combination of a hexameric helicase and a membrane ion channel (Walker 1998). However, the structures of the membrane segments of the F/V-ATPases (F_O and V_O, respectively, see Fig. 1) have little in common with membrane channels or transporters, which are usually formed by bundles of α-helices (von Heijne 2006). As shown in Fig. 1b, the c-oligomers are wide, lipid-plumbed membrane pores with internal diameters of ~3 and ~2 nm for V_O and F_O, respectively (Meier et al. 2005; Murata et al. 2005). Conceivably, such a pore (without lipid plumbing) was large enough to allow passive import and export of biopolymers in primordial cells. When combined with an ATP-driven RNA helicase, this type of membrane pore could yield an active RNA translocase that subsequently would give rise to an ATP-driven protein translocase, as depicted in Fig. 2. Then it is not surprising that a direct homologous relationship exists between the F/V-ATPases and those subunits of the bacterial flagellar motors and Type III secretion system (T3SS) that are responsible for the ATP-driven export of flagellin or secreted proteins by these

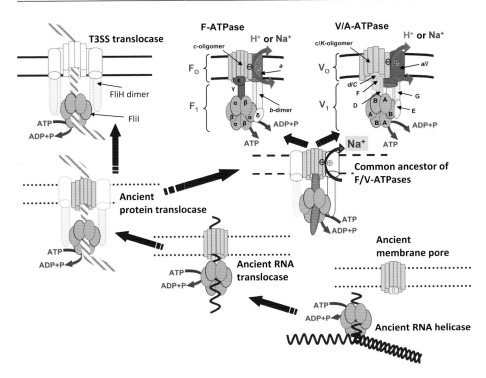

Fig. 2. **The proposed scenario of evolution from separate RNA helicases and primitive membrane pores, via membrane RNA and protein translocases, to the ion-translocating membrane ATPases.** The color code is as in Fig. 1; ancient/uncharacterized protein subunits are not colored. The striped shapes denote the translocated, partially unfolded proteins. The presence of two peripheral stalks in the primordial protein translocase and the flagellar/T3SS systems is purely hypothetical and based on the consideration that a system with one peripheral stalk would be unstable in the absence of the translocated substrate. The involvement of two FliH subunits in each peripheral stalk is based on the ability of FliH dimers to form a complex with one FliI subunit (Minamino and Namba 2004; Imada et al. 2007).

machines. This relationship can be traced through the catalytic subunits (Vogler et al. 1991) and the subunits of the peripheral stalk of the F/V-ATPases (Pallen et al. 2006).

As discussed in more detail elsewhere (Mulkidjanian et al. 2007), there is a plausible path for the transition from a protein translocase to an ion-translocating machine. The key to the transition is decrease of the pore conductivity, possibly, as a result of several amino acid replacements in the c-subunit, which would cause translocated proteins to get stuck in the translocase. Then, the torque from ATP hydrolysis, transmitted by the stuck substrate polypeptide, would cause rotation of the c-ring relative to the ex-centric membrane stator. This rotation could eventually be coupled with transmembrane ion translocation along the contact interface, via membrane-embedded, charged amino acid side chains that, otherwise, keep together the membrane subunits. Given that the structural requirements for a central stalk are likely to be minimal (Mnatsakanyan et al. 2009), this scenario naturally incorporates independent recruitment of unrelated and even structurally dissimilar proteins as central stalks in ancestral archaea and bacteria. The transition from a protein translocase to an ATP-driven ion translocase would be complete with the recruitment of the central stalk subunits, i.e., inclusion of their genes in the operons of the F-type and V-type ATPases, respectively (Mulkidjanian et al. 2007).

3 Emergence of integral membrane proteins

In the previous section, we have noted that the common ancestor of the *c*-oligomers in the F-ATPases and V-ATPases could initially function as a membrane pore. As argued by several authors (Frausto da Silva and Williams 1991; Szathmáry 2007), such pores could be needed to enable exchange of small molecules and even polymers between proto-cells and their environment. At the same time, they could represent a transition state towards the first integral membrane proteins. Integral membrane proteins contain long stretches of hydrophobic amino acid residues. By contrast, in water-soluble globular proteins, the distribution of polar and non-polar amino acids in the polypeptide chain is quasi-random (Finkelstein and Ptitsyn 2002). Assuming that the quasi-random distribution pattern is an ancestral trait, a gradual, multi-step transition from soluble proteins to membrane proteins with long hydrophobic stretches has to be envisaged. Furthermore, modern membrane proteins are co-translationally inserted into the membrane by the translocon machinery that ensures proper protein folding in the membrane (White and von Heijne 2008). The translocon itself is a membrane-bound protein complex that could not have existed before the membrane proteins evolved. In the absence of the translocon, a hydrophobic protein, if even occasionally synthesized, would remain stuck to a primeval ribosome. Therefore, a scenario of the membrane evolution must enclose an evolutionary scenario for the emergence of integral membrane proteins.

The global evolutionary analysis of integral membrane proteins by Saier et al. led to the conclusion that the evolution went from non-specific oligomeric channels, which were built of peptides with only a few transmembrane segments, towards larger, specific membrane translocators that emerged by gene duplication (Saier 2003), see also the chapter by Saier et al. in this volume. Still, the widespread notion that a stand-alone hydrophobic α-helix could, via multiple gene duplication, yield increasingly complex membrane proteins (see e.g., Popot and Engelman 2000) does not appear plausible: a solo, water-insoluble α-helix could hardly leave the ribosome in the absence of a translocon complex.

9

Physically more plausible are the scenarios that start from amphiphilic α-helices (Pohorille et al. 2003; Mulkidjanian et al. 2009). The simplest α-helical protein fold is an α-helical hairpin (*long alpha-hairpin* according to the SCOP classification (Andreeva et al. 2008). These hairpins are stabilized via hydrophobic interaction of the two α-helices. Since such stabilization is unlikely to be particularly strong, a hairpin, upon an eventual interaction with a membrane, might spread on its surface and then reassemble within the membrane in such a way that the non-polar side chains would interact with the hydrophobic lipid phase. The hairpins, then, should tend to aggregate, leading to the formation of water-filled pores, inside which the polar surfaces of α-helices would be stabilized. This arrangement seems to be partially retained by the *c*-ring of the F-ATPase that is built up of α-helical hairpins (see Fig. 1b) and is sealed by lipid only from the periplasmic side of the membrane. From the cytoplasmic site, the cavity is lined by polar residues and is apparently filled with segment(s) of the γ-subunit and water (Pogoryelov et al. 2008). The described mechanism of spontaneous protein insertion into the membrane, which does not require translocon machinery, is still used by certain bacterial toxins and related proteins. Those proteins are monomeric in their water-soluble state, but oligomerize in the membrane with the formation of pores (see Parker and Feil 2005; Anderluh and Lakey 2008, and references therein).

Membrane pores could be formed, in principle, not only by many small hairpins – which themselves could result from multiple duplication events, as inferred for the *c*-subunit of the F/V-type ATPase (Davis 2002) – but also by larger amphiphilic proteins that, after binding to membranes, might undergo "inside-out" rearrangements (see also Engelman and Zaccai 1980) with the formation of a water-filled pore in the middle of a helical bundle. This kind of protein architecture is exemplified by SecY (Van den Berg et al. 2004), another ubiquitous membrane protein besides the *c*-subunit of the F/V-ATP synthase. Starting from the pores that were built up of amphiphilic stretches of amino acids, integral membrane proteins could then evolve via the combined effect of (i) multiple replacements of polar amino acids by non-polar ones, and (ii) gene duplications, ultimately yielding multi-helix hydrophobic bundles (Saier 2000, 2003). Concomitantly, some membrane proteins would form the first translocons, enabling controlled insertion of these hydrophobic bundles into the membrane (White and von Heijne 2008).

4 Emergence of lipid membranes

The first membrane proteins required lipid membranes. What were their origins? The comparison of bacteria and archaea can hardly help to clarify the origins of lipid membranes because, as already noted, they are fundamentally different in these two domains (see Boucher et al. 2004; Pereto et al. 2004; Thomas and Rana 2007, for reviews). In both prokaryotic domains, phospholipids are built of glycerol phosphate (GP) moieties to which two hydrophobic hydrocarbon chains are attached. The GP moieties, however, are different: while bacteria use *sn*-glycerol-1-phosphate (G1P), archaea utilize its optical isomer *sn*-glycerol-3-phosphate (G3P). The hydrophobic chains, with a few exceptions, differ as well, based on fatty acids in bacteria and on isoprenoids in archaea. In bacterial lipids, the hydrophobic tails are linked to the glycerol moiety by ester bonds whereas archaeal lipids contain ether bonds. The difference extends beyond the chemical structures of the phospholipids, to the evolutionary provenance of the enzymes involved in the synthesis of phospholipids – they are either non-homologous or distantly related but not orthologous in bacteria and archaea (Boucher et al. 2004; Pereto et al. 2004; Koonin and Martin 2005; Koga and Morii 2007).

The evolutionary stage when the first lipid membranes could emerge is also uncertain. The "lipids early" models suggest that the first life forms, presumably RNA-based, were enclosed in lipid vesicles from the very beginning (see e.g., Segre et al. 2001; Deamer 2008), whereas the "lipids late" models suggest that lipid membranes could be preceded by the emergence and evolution of simple, virus-like, RNA/protein life forms (see e.g., Martin and Russell 2003; Koonin and Martin 2005; Koonin 2006).

Several lines of evidence support the "lipids late" schemes.

(a) The "lipids early" schemes imply that the first lipids were recruited from the available abiogenically synthesized compounds. Although amphiphilic molecules such as fatty acids are found in meteorites (Deamer and Pashley 1989) and could be present on the primeval Earth, it is unlikely that they all had uniformly long hydrophobic tails, which is a pre-condition for the formation of a stable bilayer. By contrast, the enzyme-synthesized amphiphilic molecules can be expected to be more homogenous.

(b) It is generally accepted that a pure lipid bilayer is not a practical solution for a primeval organism because it would prevent any exchange between the interior and the environment. Therefore, the "lipids early" models suggest that the first membranes were leaky, enabling the exchange of low-molecular compartments with the surrounding mileau (Deamer 2008). The existence of the first life forms should, however, also depend on their ability to exchange genes and to share enzymes (Koonin and Martin 2005; Szathmáry 2007). The known machines for the translocation of biological polymers across the membrane are made of proteins, which implies a co-evolution of membrane proteins and lipids.

(c) Table 1 contains the list of ubiquitous genes that are likely to be present in the LUCA. Only 2 of these ca. 60 entries, namely the above discussed *c*-subunit of the F/V-ATPases and the SecY pore subunit, belong to membrane proteins. This under-representation of membrane proteins suggests that the emergence of membrane proteins (and membranes) may have followed the emergence of RNA/protein organisms.

(d) The existence of a pre-cellular RNA/protein world is supported by the finding of viral hallmark genes shared by many groups of RNA and DNA viruses–but missing in cellular life forms. The inhabitants of this world might have been virus-like particles enclosed in protein envelopes (Koonin 2006; Koonin et al. 2006).

A really strong argument in favor of the "lipids early" models is that the lipid vesicles, by separating the first replicating entities, may have enabled their Darwinian selection (see e.g., Monnard and Deamer 2001). The primeval compartmentalization, however, could have been achieved even without lipid vesicles. Russell et al. have hypothesized that the early stages of evolution may have taken place inside iron–sulfide bubbles that formed at warm, alkaline hydrothermal vents (Russell and Hall 1997, 2006; Martin and Russell 2003). It has been suggested that iron-sulfide "bubbles" could encase LUCA consortia of small, virus-like replicating entities (Koonin and Martin 2005; Koonin et al. 2006). Such entities could share a common pool of metabolites and genes, so that each interacting consortium, e.g., inhabitants of one inorganic "bubble" at a hydrothermal vent, would comprise a distinct evolutionary unit. Such a scheme, with an extensive (gene) exchange between the members of one consortium but not between different, mechanistically separated consortia solves a major conundrum between the notion of extensive gene mixing that is considered a major feature of early evolution (Woese 1998) and the requirement of separately evolving units as subjects of Darwinian selection (Koonin and Martin 2005; Mulkidjanian et al. 2009).

This "inorganic" solution of the compartmentalization problem is further exploited in the recent "Zinc world" scenario according to which the life on Earth emerged, powered by solar radiation, within photosynthetically active precipitates of zinc sulfide (ZnS; Mulkidjanian 2009; Mulkidjanian and Galperin 2009). Honeycomb-like ZnS precipitates are widespread at the sites of deep sea hydrothermal activity (Takai et al. 2001; Hauss et al. 2005; Kormas et al. 2006; Tivey 2007). Here, the extremely hot hydrothermal fluids leach metal ions from the crust and bring them to the surface (Kelley et al. 2002; Tivey 2007). Since hydrothermal fluids are rich in H_2S, their interaction with cold ocean water leads to the precipitation of metal sulfide particles that form "smoke" over the "chimneys" of hydrothermal vents (Kelley et al. 2002; Tivey 2007). These particles eventually aggregate, settle down,

Table 1. Products of ubiquitous genes and their association with essential divalent metals (the table is taken from Mulkidjanian and Galperin 2009)

Protein function	EC number (if available)	Functional dependence on metals	Metals in at least some structures
Products of ubiquitous genes, according to Koonin (2003)			
Translation and ribosomal biogenesis			
Ribosomal proteins (33 in total)		Mg	Mg, Zn
Seryl-tRNA synthetase	6.1.1.11	Mg, Zn	Mn, Zn
Methionyl tRNA synthetase	6.1.1.10	Mg, Zn	Zn
Histidyl tRNA synthetase	6.1.1.21	Mg	No metals seen
Tryptophanyl-tRNA synthetase	6.1.1.2	Mg, Zn	Mg
Tyrosyl-tRNA synthetase	6.1.1.1	Mg	No metals seen
Phenylalanyl-tRNA synthetase	6.1.1.20	Mg, Zn	Mg
Aspartyl-tRNA synthetase	6.1.1.12	Mg	Mg, Mn
Valyl-tRNA synthetase	6.1.1.9	Mg	Zn
Isoleucyl-tRNA synthetase	6.1.1.5	Mg, Zn	Zn
Leucyl-tRNA synthetase	6.1.1.4	Mg	Zn
Threonyl-tRNA synthetase	6.1.1.2	Mg, Zn	Zn
Arginyl-tRNA synthetase	6.1.1.19	Mg	No metals seen
Prolyl-tRNA synthetase	6.1.1.15	Mg, Zn	Mg, Zn, Mn
Alanyl-tRNA synthetase	6.1.1.7	Mg, Zn	Mg, Zn
Translation elongation factor G	3.6.5.3	Mg	Mg
Translation elongation factor P/ translation initiation factor eIF5-a			Zn
Translation initiation factor 2			Zn
Translation initiation factor IF-1			No divalent metals
Pseudouridylate synthase	5.4.99.12	Mg, Zn	No metals seen
Methionine aminopeptidase	3.4.11.18	Mn, Zn, or Co	Mn or Zn or Co
Transcription			
Transcription antiterminator NusG	–	–	No metals seen
DNA-directed RNA polymerase, subunits α, β, β′	2.7.7.6	Mg	Mg, Mn, Zn
Replication			
DNA polymerase III, subunit β	2.7.7.7	Mg	Mg
Clamp loader ATPase (DNA polymerase III, subunits γ and τ)	2.7.7.7	Mg	Mg, Zn
Topoisomerase IA	5.99.1.2	Mg	No metals seen
Repair and recombination			
5′–3′ exonuclease (including N-terminal domain of PolI)	3.1.11.-	Mg	Mg
RecA/RadA recombinase	–	–	Mg

Continued on next page

Table 1. *Continued*

Protein function	EC number (if available)	Functional dependence on metals	Metals in at least some structures
Chaperone function			
Chaperonin GroEL	3.6.4.9	Mg	Mg
O-sialoglycoprotease/ apurinic endonuclease	3.4.24.57	Zn	Mg, Fe
Nucleotide and amino acid metabolism metabolism			
Thymidylate kinase	2.7.4.9	Mg	Mg
Thioredoxin reductase	1.8.1.9	–	No metals seen
Thioredoxin		–	Zn
CDP-diglyceride-synthase	2.7.7.41	Mg	No entries
Energy conversion			
Phosphomannomutase	5.4.2.8	Mg	Mg, Zn
Catalytic subunit of the membrane ATP synthase	3.6.1.34	Mg	Mg
Proteolipid subunits of the membrane ATP synthase	3.6.1.34	–	No metals seen
Triosephosphate isomerase	5.3.1.1	–	No metals seen
Coenzymes			
Glycine hydroxymethyltransferase	2.1.2.1	Mg	No metals seen
Secretion			
Preprotein translocase subunit SecY	–	–	Zn
Signal recognition particle GTPase FtsY	–	–	Mg
Miscellaneous			
Predicted GTPase	–	–	No metal ligands in the structures
Additional ubiquitous gene products from Charlebois and Doolittle (2004)			
DNA primase (dnaG)	2.7.7.7	–	Zn
S-adenosylmethionine-6-N,N′-adenosyl (rRNA) dimethyltransferase (KsgA)	2.1.1.48	Mg	No metals seen
Transcription pausing, L factor (NusA)	–	–	No metals seen

The lists of ubiquitous genes were extracted from refs. Koonin (2003) and Charlebois and Doolittle (2004). The data on the dependence of functional activity on particular metals were taken from the BRENDA Database (Chang et al. 2009). According to the BRENDA database, the enzymatic activity of most Mg-dependent enzymes could be routinely restored by Mn. As concentration of Mg^{2+} ions in the cell is ca. 10^{-2} M, whereas that of Mn^{2+} ions is ca. 10^{-6} M, the data on the functional importance of Mn were not included in the table. The presence of metals in protein structures was as listed in the Protein Data Bank (Henrick et al. 2008) entries. See Mulkidjanian and Galperin (2009) for further details and references.

and, ultimately, form sponge-like structures around the vent orifices. The sulfides of iron and copper precipitate promptly (Seewald and Seyfried 1990), their deposition starts already inside the orifices of hydrothermal vents (Kormas et al. 2006). The sulfides of zinc and manganese precipitate slower (Seewald and Seyfried 1990) and can spread over, forming halos around the iron–sulfur apexes of hydrothermal vents (see Tivey 2007, for a recent review). The Zn world model suggests that under the high pressure of the primeval, CO_2-dominated atmosphere, very hot, Zn-enriched hydrothermal fluids could reach even the sub-aerial, illuminated environments, so that ZnS could precipitate within reach of UV-rich solar beams (nowadays such hot fluids can discharge to the continental surface only as steam geysers). ZnS is a very powerful photocatalyst; it can reduce CO_2 to formate with a quantum yield of up to 80% (Henglein 1984; Henglein et al. 1984; Kanemoto et al. 1992; Eggins et al. 1993), can produce diverse other organic compounds from CO_2 (Fox and Dulay 1993; Eggins et al. 1998), including the intermediates of the Krebs cycle (Zhang et al. 2007; Guzman and Martin 2009), and can drive various transformations of carbon- and nitrogen-containing substrates (Yanagida et al. 1985; Kisch and Künneth 1991; Kisch and Lindner 2001; Marinkovic and Hoffmann 2001; Ohtani et al. 2003). In the illuminated environments, the UV light, serving as a selective factor, may have favored the accumulation of RNA-like polymers as particular photostable (Mulkidjanian et al. 2003; Sobolewski and Domcke 2006). A direct contact of the first RNA-based life forms with the surfaces of porous ZnS compartments should be of key importance: these surfaces, besides catalyzing abiogenic photosynthesis of useful metabolites and serving as templates for the synthesis of longer biopolymers from simpler building blocks, could prevent the first biopolymers from photo-dissociation by absorbing from them the excess radiation (Mulkidjanian 2009). The idea that the first RNA molecules may have been shaped by ZnS surfaces is supported by an almost perfect match of the distances that separate the positively charged Zn^{2+} ions at the ZnS surface (Dinsmore et al. 2000) with the distances between the phosphate groups in the RNA backbone (0.58–0.59 nm; Saenger 1984). In addition, Zn^{2+} ions showed an exclusive ability to catalyze the formation of naturally occurring 3′–5′ linkages upon abiogenic polymerization of nucleotides (Bridson and Orgel 1980; Van Roode and Orgel 1980).

As the ZnS-mediated photosynthesis is accompanied by the release of Zn^{2+} ions (Henglein 1984; Kisch and Künneth 1991), it should yield a steadily Zn-enriched milieu within ZnS compartments. A Zn-rich milieu is geologically unusual; the equilibrium concentration of Zn in the anoxic primeval waters was estimated as 10^{-15}–10^{-12} M (Zerkle et al. 2005; Dupont et al. 2006; Williams and Frausto da Silva 2006). If the LUCA consortia indeed dwelled within photosynthesizing ZnS compartments, then Zn^{2+} ions could be preferably recruited as metal cofactors by the proteins and RNA molecules of the LUCA. This prediction is easily testable. Table 1 exemplifies that the ubiquitous proteins – which are likely to be present in the LUCA – show notable preference for Zn as compared to other transition metals (see refs. Mulkidjanian 2009; Mulkidjanian and Galperin 2009, for further details on the Zn world scenario).

The photosynthesizing Zn world, however, could exist only as long as the pressure of the CO_2 dominated atmosphere was high enough to enable delivery of very hot, Zn-enriched hydrothermal fluids at illuminated settings. When the atmospheric pressure dropped below ca. 10 bar, the continental hydrothermal fluids should cool down and become gradually depleted of Zn ions, so that fresh ZnS surfaces could no longer form in sub-aerial settings, but only deeply at the sea floor. The organisms would have found alternative ways to reduce CO_2 and should have learned to deal with Fe^{2+}, the dominating transition metal ion in primordial sea (with an estimated content of 10^{-5}–10^{-6}M; Zerkle et al. 2005; Dupont et al. 2006; Williams and Frausto da Silva 2006). Iron, unlike zinc, can generate harmful hydroxyl radicals and is therefore detrimental for RNA (Meares et al. 2003; Cohn et al. 2004, 2006; Luther and Rickard 2005). Lipids can prevent the damaging action of iron-containing minerals on RNA (Cohn et al. 2004), so that the need to protect biopolymers from iron-containing surfaces could have prompted the transition from surface-confined replicators to lipid-encased life forms.

Why then are the lipid membranes of modern archaea and bacteria so different? Several hypotheses were suggested to explain the aforementioned usage of different GP enantiomers by archaea and bacteria. Koga has suggested that the first GP moieties were racemic because of their abiogenic origin; only later the enzymes for the synthesis of G1P and G3P separately evolved in archaea and bacteria, respectively (Koga et al. 1998; Koga and Morii 2007). Wächershäuser has suggested that membranes of pre-cells were built of lipids that contained racemic GPs units that were synthesized by a primitive non-stereospecific enzyme. The further segregation of the G1P- and G3P-containing lipids was suggested to be physico-chemical, so that lipids that carried the same GP enantiomers clustered together and eventually yielded subpopulations of organisms enriched in either enantiomeric phospholipid. It was suggested further that the higher stability of "homochiral" over "heterochiral" membranes could favor the emergence of different enzymes for stereospecific synthesis of different GP enantiomers in archaea and bacteria, respectively (Wächtershäuser 2003). Pereto et al. (2004) have hypothesized that G1P and G3P were initially synthesized in a non-specific way, as byproducts of two different dehydrogenases already present in the cenancestor, and that specific enzymes for the synthesis of G1P and G3P separately evolved from these two dehydrogenases in archaeal and bacterial lineages, respectively.

All these hypotheses are based on the assumption that the phospholipids of the LUCA (or pre-cells, or cenancestor) contained GP moieties that, as in modern membranes, linked two lipid "tails" together. In fact, there is no evidence that the very first membranes were built in this way. Even the modern membranes contain, besides GP-containing two-tailed phospholipids, also single-tailed fatty acids and four-tailed cardiolipin molecules. The concept of gradual, multistep membrane evolution, as outlined in previous sections, is better compatible with a scenario where the first lipids could be simple and single-tailed. As argued above, the function of first, supposedly porous, membranes was limited to occluding biological polymers while enabling the exchange of small molecules and ions. The experiments with simple amphiphilic compounds have shown that vesicles made either of fatty acids (Deamer and Dworkin 2005; Deamer 2008) or of phosphorylated isoprenoids (Nomura et al. 2002; Gotoh et al. 2006; Streiff et al. 2007) can entrap polynucleotides and proteins. Isoprenoids were likely to be present at the stage of LUCA: their enzymatic synthesis is simple, and they are found in all domains of life, unlike fatty acids that, most likely, have emerged in the bacterial lineage (Smit and Mushegian 2000). Hence, one can speculate that the leaky membranes of LUCA were simple, being built of e.g., phosphorylated isoprenoids. To attain ion-tight membranes, the first cells, however, had to stabilize the membrane/water interface and increase the thickness of the membrane, since the permeability of lipid bilayer to small ions (with exception of protons, see above) is limited by ion penetration across the membrane/water interface and depends on the membrane thickness (Deamer 1987; Nagle 1987; Tepper and Voth 2006). A pair-wise linking of hydrophobic tails by GP moieties seems to be the chemically simplest way to solve both tasks: the membrane interface becomes less leaky to ions and the thickness of the bilayer increases by ca. 0.6 nm. In addition, the phosphate moiety of GP ensures the amphiphilicity of the bilayer and an eventual binding of a head group. Bacteria and archaea may have found this simple solution independently, by using different GP enantiomers and unrelated enzymes. In Bacteria this transition may have been accompanied by the recruitment of fatty acids; the isoprenoid derivatives, however, were retained by bacterial membranes, in particular, as hopanoids and single-tailed quinones (Haines 2001; Hauss et al. 2005).

5 Scenario for the origin and evolution of membranes and membrane proteins

Apparently, the central theme in the early cellular evolution was the increasing tightness of cell envelopes. Indeed, the emergence of such a complex device, that is the modern biological membrane, could proceed only via many intermediate stages. Szathmáry et al. have recently developed and modeled a set of evolutionary scenarios that exemplified the crucial importance of the interaction and exchange between the primeval replicating entities for the stability of their populations (Szathmáry 2006, 2007; Könnyü et al. 2008; Branciamore et al. 2009). According to Szathmáry

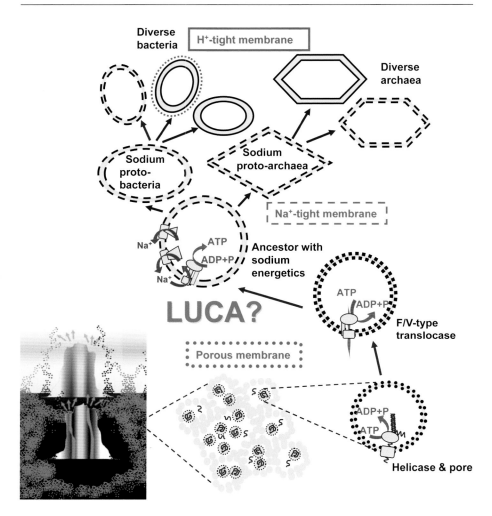

Fig. 3. The proposed scenario for the evolution of membranes and membrane enzymes. The scheme suggests the emergence of first replicating entities within honeycomb-like ZnS precipitates of hydrothermal origin. Note that FeS and ZnS particles (black and gray dots, respectively) precipitate at different distances from the hot spring (the picture is based on data from Seewald and Seyfried 1990; Takai et al. 2001; Kelley et al. 2002; Kormas et al. 2006; Russell 2006). The evolution of membranes is shown as a transition from primitive, porous membranes that were leaky both to Na$^+$ and H$^+$ (dotted lines), via membranes that were Na$^+$-tight but H$^+$-leaky (dashed lines) to the modern-type membranes that are impermeable to both H$^+$ and Na$^+$ (solid lines). As the common ancestor of the F- and V-ATPases possessed a Na$^+$-binding site (Mulkidjanian et al. 2008b, 2009), the LUCA (regardless of whether it was a modern-type cell or a consortium that included replicating, membrane-surrounded entities) either had porous membranes so that the common ancestor of the F- and A/V-ATPases operated as a polymer translocase, with Na$^+$ ions performing a structural role, or had membranes that were tight to sodium ions but permeable to protons; in this case the LUCA could possess sodium energetic (see main text, and Mulkidjanian et al. 2008b, 2009; Mulkidjanian 2009; Mulkidjanian and Galperin 2009, for details).

18

(2007), an increase in the complexity of pro-cells should be accompanied by their progressive sequestering from the environment, so that the gradual build-up of enzymatic pathways inside the pro-cells would be accompanied by decrease in membrane permeability. Figure 3 depicts a tentative scenario of a co-evolution of membranes and membrane proteins where the gradual decrease in membrane permeability, on the one hand, enables the emergence of new enzyme systems that demand tight membranes and, on the other hand, leads to expunction of "leaky" membrane proteins.

The scenario starts from simple replicating entities that may have dwelled in honeycomb-like mineral compartments, which, in the framework of the Zn world scenario (Mulkidjanian 2009; Mulkidjanian and Galperin 2009), could help to (photo) select the first RNA organisms, provide a shelter and nourish them. At this stage, the first replicators could survive only by sharing metabolites, enzymes, and genes. Gradually, however, the first life forms may have attained protecting envelopes that initially could be built predominantly of proteins. The subsequent transition to the predominantly lipid membranes should be accompanied by the emergence of primitive membrane pores that might resemble the c-rings of the F-type and V-type ATPases. The requirement for horizontal gene transfer and gene mixing should, however, drive the emergence of active, ATP-driven RNA and protein translocases, giving rise, in particular, to the ancestor of the F/V-type ATPases, which, apparently, was a chimera of a (former) RNA-helicase and a membrane pore.

The next stage of evolution is envisaged as selection for tighter membranes that would maintain the ionic homeostasis of the evolving cells. According to the principle of chemistry conservation (see e.g., Mulkidjanian and Galperin 2007), primordial cells would strive to keep their internal chemistry similar to the chemical compositions of the brine in which the first life forms had emerged. Besides the need to maintain a high internal Zn concentration (after the supposed dramatic shift of the Zn/Fe ratio in their habitats, see also Mulkidjanian and Galperin 2009), the first cells should be also challenged by growing sodium content in the sea water. Since the cytoplasm of all cells contains more potassium than sodium, and the translation systems specifically require K^+ for functioning (Bayley and Kushner 1964; Spirin et al. 1988), the first life forms were likely to emerge in K^+-rich environments (Natochin 2007; Mulkidjanian 2009). The concentration of Na^+ in the sea water, should, however, increase with time (DeRonde et al. 1997; Foriel et al. 2004; Pinti 2005), affecting the Na^+/K^+ ratio inside the pro-cells. These challenges should strongly favor evolution both of ion-tight membranes and of ion pumps, in particular those capable of expunging Na^+ ions out of the cell. This requirement could be behind the transition from a protein translocase to the precursor of a Na^+-translocating membrane ATPase.

Most likely, the ancestral rotating ATPases would pump Na$^+$ along with other Na$^+$-pumps, such as the Na$^+$-transporting pyrophosphatase (Malinen et al. 2007) and Na$^+$-transporting decarboxylase (Dimroth 1997), which are present in both bacteria and archaea and appear to antedate the divergence of the three domains of life. Unlike the other Na$^+$ pumps, the common ancestor of the V/F-ATPases, owing to its rotating scaffold, would be potentially able to translocate Na$^+$ ions in both directions. Upon further increase in the ocean salinity, reversal of the rotation would result in Na$^+$-driven synthesis of ATP by this primordial rotary machine. Already in Archaean, the concentration of Na$^+$ in the ocean water was approx. 1 M (DeRonde et al. 1997; Foriel et al. 2004; Pinti 2005), i.e., it was high enough for the rotary machine to switch from the ATP hydrolysis to the ATP synthesis mode. This event marked the birth of membrane bioenergetics: together with the ancient Na$^+$ pumps, the ancestral V/F-type ATP synthases would complete the first, sodium-dependent bioenergetic cycle in a cell membrane, as shown in Fig. 3.

The final evolutionary step in the present scenario is envisaged as transition to proton-tight, elaborate membranes that provided better protection to the cells. These membranes, in addition, were more lucrative from the point of view of energetics: proton transfer can be chemically coupled to redox reactions, especially those of water and diverse quinones, thus enabling the advent of efficient redox- and light-driven generators of PMF, such as cytochrome bc_1 complex (Mulkidjanian 2007), cytochrome oxidase (Brzezinski 2004) or water-splitting photosystem II (Junge et al. 2002). Therefore, once the membranes could maintain PMF and the first proton pumps emerged, the sodium-binding sites of the F-type and V-type ATPases became obsolete and deteriorated independently in multiple lineages. Ancestral, less effective sodium bioenergetics persisted in anaerobic thermophiles and alkaliphiles that cannot benefit from proton energetics and in some marine and parasitic bacteria and archaea that exist in high-sodium environments (Mulkidjanian et al. 2008a). Further traces of Na$^+$-based bioenergetics are seen in the universal distribution of Na$^+$ gradients and Na$^+$-dependent systems of solute transport in virtually all known cell types. In particular, plasma membranes of animal cells remained proton-leaky, "sodium membranes" (Skulachev 1988) inasmuch as they, although with some exceptions (Wieczorek et al. 1999), cannot maintain H$^+$ gradient.

In conclusion, we would like to submit that the evolution of membrane proteins should be considered together with the evolution of the membrane lipids since the specific physical properties of lipid bilayers, in particular, their permeability, control the functions that membrane proteins can perform.

Acknowledgments

We are grateful to Eugene Koonin for numerous discussions, which shaped our views on the evolution of first organisms. Valuable suggestions from Dr. A.A. Baykov, Dr. P.A. Dibrov, Dr. G. Fox, Dr. T. Haines, Dr. K.S. Makarova, Dr. T. Meier, Dr. D.A. Pogoryelov, Dr. V.P. Skulachev, Dr. G. Wächtershäuser, Dr. J.E. Walker, and Dr. Y.I. Wolf are greatly appreciated. This study was supported by grants from the Deutsche Forschungsgemeinschaft and the Volkswagen Foundation (A.Y.M.), and by the Intramural Research Program of the National Library of Medicine at the National Institutes of Health (M.Y.G.).

References

Anderluh G and Lakey JH (2008) Disparate proteins use similar architectures to damage membranes. Trends Biochem Sci 33: 482–490

Andreeva A, Howorth D, Chandonia JM, Brenner SE, Hubbard TJ, Chothia C, Murzin AG (2008) Data growth and its impact on the SCOP database: new developments. Nucleic Acids Res 36: D419–D425

Angevine CM, Herold KA, Fillingame RH (2003) Aqueous access pathways in subunit a of rotary ATP synthase extend to both sides of the membrane. Proc Natl Acad Sci USA 100: 13179–13183

Bayley ST and Kushner DJ (1964) The ribosomes of the extremely halophilic bacterium, *Halobacterium cutirubrum*. J Mol Biol 9: 654–669

Beyenbach KW and Wieczorek H (2006) The V-type H+ ATPase: molecular structure and function, physiological roles and regulation. J Exp Biol 209: 577–589

Boucher Y, Kamekura M, Doolittle WF (2004) Origins and evolution of isoprenoid lipid biosynthesis in archaea. Mol Microbiol 52: 515–527

Boyer PD (1997) The ATP synthase – a splendid molecular machine. Annu Rev Biochem 66: 717–749

Branciamore S, Gallori E, Szathmáry E, Czaran T (2009) The origin of life: chemical evolution of a metabolic system in a mineral honeycomb? J Mol Evol 69: 458–469

Bridson PK and Orgel LE (1980) Catalysis of accurate poly(C)-directed synthesis of 3′–5′-linked oligoguanylates by Zn^{2+}. J Mol Biol 144: 567–577

Brzezinski P (2004) Redox-driven membrane-bound proton pumps. Trends Biochem Sci 29: 380–387

Cao TB and Saier MH Jr (2003) The general protein secretory pathway: phylogenetic analyses leading to evolutionary conclusions. Biochim Biophys Acta 1609: 115–125

Capaldi RA and Aggeler R (2002) Mechanism of the F_1F_0-type ATP synthase, a biological rotary motor. Trends Biochem Sci 27: 154–160.

Chang A, Scheer M, Grote A, Schomburg I, Schomburg D (2009) BRENDA, AMENDA and FRENDA the enzyme information system: new content and tools in 2009. Nucleic Acids Res 37: D588–D592

Charlebois RL and Doolittle WF (2004) Computing prokaryotic gene ubiquity: rescuing the core from extinction. Genome Res 14: 2469–2477

Cherepanov DA, Mulkidjanian AY, Junge W (1999) Transient accumulation of elastic energy in proton translocating ATP synthase. FEBS Lett 449: 1–6

Cohn CA, Borda MJ, Schoonen MA (2004) RNA decomposition by pyrite-induced radicals and possible role of lipids during the emergence of life. Earth Planet Sci Lett 225: 271–278

Cohn CA, Mueller S, Wimmer E, Leifer N, Greenbaum S, Strongin DR, Schoonen MA (2006) Pyrite-induced hydroxyl radical formation and its effect on nucleic acids. Geochem Trans 7: 3

Cramer WA and Knaff DB (1990) Energy transduction in biological membranes: a textbook of bioenergetics. Springer, New York

Davis BK (2002) Molecular evolution before the origin of species. Prog Biophys Mol Biol 79: 77–133

Deamer DW (1987) Proton permeation of lipid bilayers. J Bioenerg Biomembr 19: 457–479

Deamer DW (1997) The first living systems: a bioenergetic perspective. Microbiol Mol Biol Rev 61: 239–261

Deamer DW (2008) Origins of life: how leaky were primitive cells? Nature 454: 37–38

Deamer DW and Pashley RM (1989) Amphiphilic components of the Murchison carbonaceous chondrite: surface properties and membrane formation. Orig Life Evol Biosph 19: 21–38

Deamer DW and Dworkin JP (2005) Chemistry and physics of primitive membranes. Top Curr Chem 259: 1–27

Deckers-Hebestreit G, Greie J, Stalz W, Altendorf K (2000) The ATP synthase of *Escherichia coli*: structure and function of F0 subunits. Biochim Biophys Acta 1458: 364–373

DeRonde CEJ, Channer DMD, Faure K, Bray CJ, Spooner ETC (1997) Fluid chemistry of Archean seafloor hydrothermal vents: implications for the composition of circa 3.2 Ga seawater. Geochim Cosmochim Acta 61: 4025–4042

Dibrov PA (1991) The role of sodium ion transport in *Escherichia coli* energetics. Biochim Biophys Acta 1056: 209–224

Dimroth P (1997) Primary sodium ion translocating enzymes. Biochim Biophys Acta 1318: 11–51

Dimroth P, von Ballmoos C, Meier T (2006) Catalytic and mechanical cycles in F-ATP synthases. EMBO Rep 7: 276–282

Dinsmore AD, Hsu DS, Qadri SB, Cross JO, Kennedy TA, Gray HF, Ratna BR (2000) Structure and luminescence of annealed nanoparticles of ZnS:Mn. J Appl Phys 88: 4985–4993

Drory O and Nelson N (2006) The emerging structure of vacuolar ATPases. Physiology (Bethesda) 21: 317–325

Dupont CL, Yang S, Palenik B, Bourne PE (2006) Modern proteomes contain putative imprints of ancient shifts in trace metal geochemistry. Proc Natl Acad Sci USA 103: 17822–17827

Dzioba J, Hase CC, Gosink K, Galperin MY, Dibrov P (2003) Experimental verification of a sequence-based prediction: F_1F_0-type ATPase of *Vibrio cholerae* transports protons, not Na^+ ions. J Bacteriol 185: 674–678

Eggins BR, Robertson PKJ, Stewart JH, Woods E (1993) Photoreduction of carbon dioxide on zinc sulfide to give four-carbon and two-carbon acids. J Chem Soc Chem Commun: 349–350

Eggins BR, Robertson PKJ, Murphy EP, Woods E, Irvine JTS (1998) Factors affecting the photoelectrochemical fixation of carbon dioxide with semiconductor colloids. J Photochem Photobiol A Chem 118: 31–40

Engelman DM and Zaccai G (1980) Bacteriorhodopsin is an inside-out protein. Proc Natl Acad Sci USA 77: 5894–5898

Esteban O, Bernal RA, Donohoe M, Videler H, Sharon M, Robinson CV, Stock D (2008) Stoichiometry and localization of the stator subunits E and G in *Thermus thermophilus* H^+-ATPase/synthase. J Biol Chem 283: 2595–2603

22

Finkelstein AV and Ptitsyn OB (2002) Protein physics: a course of lectures. Academic Press, Amsterdam

Forgac M (2007) Vacuolar ATPases: rotary proton pumps in physiology and pathophysiology. Nat Rev Mol Cell Biol 8: 917–929

Foriel J, Philippot P, Rey P, Somogyi A, Banks D, Menez B (2004) Biological control of Cl/Br and low sulfate concentration in a 3.5-Gyr-old seawater from North Pole, Western Australia. Earth Planet. Sci Lett 228: 451–463

Fox MA and Dulay MT (1993) Heterogeneous photocatalysis. Chem Rev 93: 341–357

Frausto da Silva JJR and Williams RJP (1991) The biological chemistry of the elements: the inorganic chemistry of life. Clarendon Press, Oxford

Gennis RB (1989) Biomembranes: molecular structure and function. Springer, New York

Gibbons C, Montgomery MG, Leslie AG, Walker JE (2000) The structure of the central stalk in bovine F_1-ATPase at 2.4 A resolution. Nat Struct Biol 7: 1055–1061

Glansdorff N, Xu Y, Labedan B (2008) The Last Universal Common Ancestor: emergence, constitution and genetic legacy of an elusive forerunner. Biol Direct 3: 29

Gogarten JP, Kibak H, Dittrich P, Taiz L, Bowman EJ, Bowman BJ, Manolson MF, Poole RJ, Date T, Oshima T, et al. (1989) Evolution of the vacuolar H+-ATPase: implications for the origin of eukaryotes. Proc Natl Acad Sci USA 86: 6661–6665

Gogarten JP, Starke T, Kibak H, Fishman J, Taiz L (1992) Evolution and isoforms of V-ATPase subunits. J Exp Biol 172: 137–147

Gotoh M, Miki A, Nagano H, Ribeiro N, Elhabiri M, Gurnienna-Kontecka E, Albrecht-Gary AM, Schmutz M, Ourisson G, Nakatani Y (2006) Membrane properties of branched polyprenyl phosphates, postulated as primitive membrane constituents. Chem Biodivers 3: 434–455

Guzman MI and Martin ST (2009) Prebiotic metabolism: production by mineral photoelectrochemistry of alpha-ketocarboxylic acids in the reductive tricarboxylic acid cycle. Astrobiology 9: 833–842

Haines TH (2001) Do sterols reduce proton and sodium leaks through lipid bilayers? Prog Lipid Res 40: 299–324

Harris JK, Kelley ST, Spiegelman GB, Pace NR (2003) The genetic core of the universal ancestor. Genome Res 13: 407–412

Häse CC, Fedorova ND, Galperin MY, Dibrov PA (2001) Sodium ion cycle in bacterial pathogens: evidence from cross-genome comparisons. Microbiol Mol Biol Rev 65: 353–370

Hauss T, Dante S, Haines TH, Dencher NA (2005) Localization of coenzyme Q10 in the center of a deuterated lipid membrane by neutron diffraction. Biochim Biophys Acta 1710: 57–62

Henglein A (1984) Catalysis of photochemical reactions by colloidal semiconductors. Pure Appl Chem 56: 1215–1224

Henglein A, Gutierrez M, Fischer CH (1984) Photochemistry of colloidal metal sulfides. 6. Kinetics of interfacial reactions at ZnS particles. Ber Der Bunsen-Ges-Phys Chem Chem Phys 88: 170–175

Henrick K, Feng Z, Bluhm WF, Dimitropoulos D, Doreleijers JF, Dutta S, Flippen-Anderson JL, Ionides J, Kamada C, Krissinel E, Lawson CL, Markley JL, Nakamura H, Newman R, Shimizu Y, Swaminathan J, Velankar S, Ory J, Ulrich EL, Vranken W, Westbrook J, Yamashita R, Yang H, Young J, Yousufuddin M, Berman HM (2008) Remediation of the protein data bank archive. Nucleic Acids Res 36: D426–D433

Hilario E and Gogarten JP (1993) Horizontal transfer of ATPase genes – the tree of life becomes a net of life. Biosystems 31: 111–119

Hilario E and Gogarten JP (1998) The prokaryote-to-eukaryote transition reflected in the evolution of the V/F/A-ATPase catalytic and proteolipid subunits. J Mol Evol 46: 703–715

Imada K, Minamino T, Tahara A, Namba K (2007) Structural similarity between the flagellar type III ATPase FliI and F_1-ATPase subunits. Proc Natl Acad Sci USA 104: 485–490

Imamura H, Takeda M, Funamoto S, Shimabukuro K, Yoshida M, Yokoyama K (2005) Rotation scheme of V_1-motor is different from that of F_1-motor. Proc Natl Acad Sci USA 102: 17929–17933

Itoh H, Takahashi A, Adachi K, Noji H, Yasuda R, Yoshida M, Kinosita K (2004) Mechanically driven ATP synthesis by F_1-ATPase. Nature 427: 465–468

Jekely G (2006) Did the last common ancestor have a biological membrane? Biol Direct 1: 35

Junge W, Haumann M, Ahlbrink R, Mulkidjanian A, Clausen J (2002) Electrostatics and proton transfer in photosynthetic water oxidation. Philos Trans R Soc Lond B Biol Sci 357: 1407–1417; discussion 1417–1420

Kanemoto M, Shiragami T, Pac CJ, Yanagida S (1992) Semiconductor photocatalysis – effective photoreduction of carbon-dioxide catalyzed by ZnS quantum crystallites with low-density of surface-defects. J Phys Chem 96: 3521–3526

Kawano M, Igarashi K, Yamato I, Kakinuma Y (2002) Arginine residue at position 573 in *Enterococcus hirae* vacuolar-type ATPase NtpI subunit plays a crucial role in Na^+ translocation. J Biol Chem 277: 24405–24410

Kawasaki-Nishi S, Nishi T, Forgac M (2001) Arg-735 of the 100-kDa subunit a of the yeast V-ATPase is essential for proton translocation. Proc Natl Acad Sci USA 98: 12397–12402

Kelley DS, Baross JA, Delaney JR (2002) Volcanoes, fluids, and life at mid-ocean ridge spreading centers. Annu Rev Earth Planet Sci 30: 385–491

Kisch H and Künneth R (1991) Photocatalysis by semiconductor powders: preparative and mechanistic aspects. In: Rabek J (ed) Photochemistry and photophysics. CRC Press Inc., Boca Raton, pp 131–175

Kisch H and Lindner W (2001) Syntheses via semiconductor photocatalysis. Chem Unserer Zeit 35: 250–257

Kitagawa N, Mazon H, Heck AJ, Wilkens S (2008) Stoichiometry of the peripheral stalk subunits E and G of yeast V_1-ATPase determined by mass spectrometry. J Biol Chem 283: 3329–3337

Koga Y and Morii H (2007) Biosynthesis of ether-type polar lipids in archaea and evolutionary considerations. Microbiol Mol Biol Rev 71: 97–120

Koga Y, Kyuragi T, Nishihara M, Sone N (1998) Did archaeal and bacterial cells arise independently from noncellular precursors? A hypothesis stating that the advent of membrane phospholipid with enantiomeric glycerophosphate backbones caused the separation of the two lines of descent. J Mol Evol 46: 54–63

Konings WN (2006) Microbial transport: adaptations to natural environments. Antonie Van Leeuwenhoek 90: 325–342

Konings WN, Albers SV, Koning S, Driessen AJ (2002) The cell membrane plays a crucial role in survival of bacteria and archaea in extreme environments. Antonie Van Leeuwenhoek 81: 61–72

Könnyü B, Czaran T, Szathmáry E (2008) Prebiotic replicase evolution in a surface-bound metabolic system: parasites as a source of adaptive evolution. BMC Evol Biol 8: 267

Koonin EV (2003) Comparative genomics, minimal gene-sets and the last universal common ancestor. Nat Rev Microbiol 1: 127–136

Koonin EV (2006) On the origin of cells and viruses: a comparative-genomic perspective. Isr J Ecol Evol 52: 299–318

Koonin EV and Martin W (2005) On the origin of genomes and cells within inorganic compartments. Trends Genet 21: 647–654

Koonin EV, Senkevich TG, Dolja VV (2006) The ancient virus world and evolution of cells. Biol Direct 1: 29

Kormas KA, Tivey MK, Von Damm K, Teske A (2006) Bacterial and archaeal phylotypes associated with distinct mineralogical layers of a white smoker spire from a deep-sea hydrothermal vent site (9 degrees N, East Pacific Rise). Environ Microbiol 8: 909–920

Lapierre P, Shial R, Gogarten JP (2006) Distribution of F- and A/V-type ATPases in *Thermus scotoductus* and other closely related species. Syst Appl Microbiol 29: 15–23

Luther GW and Rickard DT (2005) Metal sulfide cluster complexes and their biogeochemical importance in the environment. J Nanoparticle Res 7: 389–407

Malinen AM, Belogurov GA, Baykov AA, Lahti R (2007) Na^+-pyrophosphatase: a novel primary sodium pump. Biochemistry 46: 8872–8878

Marinkovic S and Hoffmann N (2001) Efficient radical addition of tertiary amines to electron-deficient alkenes using semiconductors as photochemical sensitisers. Chem Commun (Camb): 1576–1577

Martin W and Russell MJ (2003) On the origins of cells: a hypothesis for the evolutionary transitions from abiotic geochemistry to chemoautotrophic prokaryotes, and from prokaryotes to nucleated cells. Philos Trans R Soc Lond B Biol Sci 358: 59–83

Meares CF, Datwyler SA, Schmidt BD, Owens J, Ishihama A (2003) Principles and methods of affinity cleavage in studying transcription. Meth Enzymol 371: 82–106

Meier T, Polzer P, Diederichs K, Welte W, Dimroth P (2005) Structure of the rotor ring of F-type Na^+-ATPase from *Ilyobacter tartaricus*. Science 308: 659–662

Meier T, Krah A, Bond PJ, Pogoryelov D, Diederichs K, Faraldo-Gomez JD (2009) Complete ion-coordination structure in the rotor ring of Na+-dependent F-ATP synthases. J Mol Biol 391: 498–507

Minamino T and Namba K (2004) Self-assembly and type III protein export of the bacterial flagellum. J Mol Microbiol Biotechnol 7: 5-17

Mnatsakanyan N, Hook JA, Quisenberry L, Weber J (2009) ATP synthase with its gamma subunit reduced to the N-terminal helix can still catalyze ATP synthesis. J Biol Chem 284: 26519–26525

Monnard PA and Deamer DW (2001) Nutrient uptake by protocells: a liposome model system. Orig Life Evol Biosph 31: 147–155

Mulkidjanian AY (2006) Proton in the well and through the desolvation barrier. Biochim Biophys Acta 1757: 415–427

Mulkidjanian AY (2007) Proton translocation by the cytochrome bc_1 complexes of phototrophic bacteria: introducing the activated Q-cycle. Photochem Photobiol Sci 6: 19–34

Mulkidjanian AY (2009) On the origin of life in the Zinc World: 1. Photosynthetic, porous edifices built of hydrothermally precipitated zinc sulfide (ZnS) as cradles of life on Earth. Biol Direct 4: 26

Mulkidjanian AY and Galperin MY (2007) Physico-chemical and evolutionary constraints for the formation and selection of first biopolymers: towards the consensus paradigm of the abiogenic origin of life. Chem Biodivers 4: 2003–2015

Mulkidjanian AY and Galperin MY (2009) On the origin of life in the Zinc World. 2. Validation of the hypothesis on the photosynthesizing zinc sulfide edifices as cradles of life on Earth. Biol Direct 4: 27

Mulkidjanian AY, Cherepanov DA, Galperin MY (2003) Survival of the fittest before the beginning of life: selection of the first oligonucleotide-like polymers by UV light. BMC Evol Biol 3: 12

Mulkidjanian AY, Makarova KS, Galperin MY, Koonin EV (2007) Inventing the dynamo machine: the evolution of the F-type and V-type ATPases. Nat Rev Microbiol 5: 892–899

Mulkidjanian AY, Dibrov P, Galperin MY (2008a) The past and present of sodium energetics: may the sodium-motive force be with you. Biochim Biophys Acta 1777: 985–992

Mulkidjanian AY, Galperin MY, Makarova KS, Wolf YI, Koonin EV (2008b) Evolutionary primacy of sodium bioenergetics. Biol Direct 3: 13

Mulkidjanian AY, Galperin MY, Koonin EV (2009) Co-evolution of primordial membranes and membrane proteins. Trends Biochem Sci 34: 206–215

Müller V and Gruber G (2003) ATP synthases: structure, function and evolution of unique energy converters. Cell Mol Life Sci 60: 474–494

Murata T, Yamato I, Kakinuma Y, Leslie AG, Walker JE (2005) Structure of the rotor of the V-type Na$^+$-ATPase from *Enterococcus hirae*. Science 308: 654–659

Mushegian A (2008) Gene content of LUCA, the last universal common ancestor. Front Biosci 13: 4657–4666

Nagle JF (1987) Theory of passive proton conductance in lipid bilayers. J Bioenerg Biomembr 19: 413–426

Nakanishi-Matsui M and Futai M (2006) Stochastic proton pumping ATPases: from single molecules to diverse physiological roles. IUBMB Life 58: 318–322

Natochin YV (2007) The physiological evolution of animals: Sodium is the clue to resolving contradictions. Herald Russ Acad Sci 77: 581–591

Nelson N (1989) Structure, molecular genetics, and evolution of vacuolar H$^+$-ATPases. J Bioenerg Biomembr 21: 553–571

Noji H, Yasuda R, Yoshida M, Kinosita K Jr (1997) Direct observation of the rotation of F$_1$-ATPase. Nature 386: 299–302

Nomura SIM, Tsumoto K, Yoshikawa K, Ourisson G, Nakatani Y (2002) Towards proto-cells: "Primitive" lipid vesicles encapsulating giant DNA and its histone complex. Cell Mol Biol Lett 7: 245–246

Ohtani B, Pal B, Ikeda S (2003) Photocatalytic organic syntheses: selective cyclization of amino acids in aqueous suspensions. Catalysis Surv Asia 7: 165–176

Pallen MJ, Bailey CM, Beatson SA (2006) Evolutionary links between FliH/YscL-like proteins from bacterial type III secretion systems and second-stalk components of the F$_O$F$_1$ and vacuolar ATPases. Protein Sci 15: 935–941

Panke O, Gumbiowski K, Junge W, Engelbrecht S (2000) F-ATPase: specific observation of the rotating c subunit oligomer of EF$_O$EF$_1$. FEBS Lett 472: 34–38

Parker MW and Feil SC (2005) Pore-forming protein toxins: from structure to function. Prog Biophys Mol Biol 88: 91–142

Patel SS and Picha KM (2000) Structure and function of hexameric helicases. Annu Rev Biochem 69: 651–697

Pereto J, Lopez-Garcia P, Moreira D (2004) Ancestral lipid biosynthesis and early membrane evolution. Trends Biochem Sci 29: 469–477

Perzov N, Padler-Karavani V, Nelson H, Nelson N (2001) Features of V-ATPases that distinguish them from F-ATPases. FEBS Lett 504: 223–228

Pinti DL (2005) The origin and evolution of the oceans. In: Gargaud M, Barbier B, Martin H, Reisse J (eds) Lectures in astrobiology. Springer, Berlin, pp 83–111

Pogoryelov D, Yu J, Meier T, Vonck J, Dimroth P, Muller DJ (2005) The c15 ring of the *Spirulina platensis* F-ATP synthase: F$_1$/F$_O$ symmetry mismatch is not obligatory. EMBO Rep 6: 1040–1044

Pogoryelov D, Nikolaev Y, Schlattner U, Pervushin K, Dimroth P, Meier T (2008) Probing the rotor subunit interface of the ATP synthase from *Ilyobacter tartaricus*. FEBS J 275: 4850–4862

Pohorille A, Wilson MA, Chipot C (2003) Membrane peptides and their role in protobiological evolution. Orig Life Evol Biosph 33: 173–197

Popot JL and Engelman DM (2000) Helical membrane protein folding, stability, and evolution. Annu Rev Biochem 69: 881–922

Rahlfs S and Müller V (1997) Sequence of subunit c of the Na$^+$-translocating F$_1$F$_0$ ATPase of *Acetobacterium woodii*: proposal for determinants of Na$^+$ specificity as revealed by sequence comparisons. FEBS Lett 404: 269–271

Russell M (2006) First Life. Am Sci 94: 32–39

Russell MJ and Hall AJ (1997) The emergence of life from iron monosulphide bubbles at a submarine hydrothermal redox and pH front. J Geol Soc London 154: 377–402

Russell M and Hall AJ (2006) The onset and early evolution of life. In: Kesler SE, Ohmoto H (eds) Evolution of early earth's atmosphere, hydrosphere, and biosphere – constraints from ore deposits, Vol. 198. Geological Society of America Memoir, pp 1–32

Saenger W (1984) Principles of nucleic acid structure. Springer Verlag, Berlin

Saier MH Jr (2000) A functional–phylogenetic classification system for transmembrane solute transporters. Microbiol Mol Biol Rev 64: 354–411

Saier MH Jr (2003) Tracing pathways of transport protein evolution. Mol Microbiol 48: 1145–1156

Seewald JS and Seyfried WE (1990) The effect of temperature on metal mobility in subseafloor hydrothermal systems: constraints from basalt alteration experiments. Earth Planet Sci Lett 101: 388–403

Segre D, Ben-Eli D, Deamer DW, Lancet D (2001) The lipid world. Orig Life Evol Biosph 31: 119–145

Skulachev VP (1988) Membrane bioenergetics. Springer-Verlag, Berlin

Smit A and Mushegian A (2000) Biosynthesis of isoprenoids via mevalonate in Archaea: the lost pathway. Genome Res 10: 1468–1484

Sobolewski AL and Domcke W (2006) The chemical physics of the photostability of life. Europhys News 37: 20–23

Spirin AS, Baranov VI, Ryabova LA, Ovodov SY, Alakhov YB (1988) A continuous cell-free translation system capable of producing polypeptides in high yield. Science 242: 1162–1164

Stock D, Gibbons C, Arechaga I, Leslie AG, Walker JE (2000) The rotary mechanism of ATP synthase. Curr Opin Struct Biol 10: 672–679

Streiff S, Ribeiro N, Wu Z, Gumienna-Kontecka E, Elhabiri M, Albrecht-Gary AM, Ourisson G, Nakatani Y (2007) "Primitive" membrane from polyprenyl phosphates and polyprenyl alcohols. Chem Biol 14: 313–319

Supekova L, Supek F, Nelson N (1995) The *Saccharomyces cerevisiae* VMA10 is an intron-containing gene encoding a novel 13-kDa subunit of vacuolar H$^+$-ATPase. J Biol Chem 270: 13726–13732

Szathmáry E (2006) The origin of replicators and reproducers. Philos Trans R Soc Lond B Biol Sci 361: 1761–1776

Szathmáry E (2007) Coevolution of metabolic networks and membranes: the scenario of progressive sequestration. Philos. Trans R Soc Lond B Biol Sci 362: 1781–1787

Takai K, Komatsu T, Inagaki F, Horikoshi K (2001) Distribution of archaea in a black smoker chimney structure. Appl Environ Microbiol 67: 3618–3629

Tepper HL and Voth GA (2006) Mechanisms of passive ion permeation through lipid bilayers: insights from simulations. J Phys Chem B 110: 21327–21337

Thomas JA and Rana FR (2007) The influence of environmental conditions, lipid composition, and phase behavior on the origin of cell membranes. Orig Life Evol Biosph 37: 267–285

Tivey MK (2007) Generation of seafloor hydrothermal vent fluids and associated mineral deposits. Oceanography 20: 50–65

Van den Berg B, Clemons WM Jr, Collinson I, Modis Y, Hartmann E, Harrison SC, Rapoport TA (2004) X-ray structure of a protein-conducting channel. Nature 427: 36–44

Van Roode JHG and Orgel LE (1980) Template-directed synthesis of oligoguanylates in the presence of metal-ions. J Mol Biol 144: 579–585

Vogler AP, Homma M, Irikura VM, Macnab RM (1991) *Salmonella typhimurium* mutants defective in flagellar filament regrowth and sequence similarity of FliI to F_0F_1, vacuolar, and archaebacterial ATPase subunits. J Bacteriol 173: 3564–3572

von Ballmoos C and Dimroth P (2007) Two distinct proton binding sites in the ATP synthase family. Biochemistry 46: 11800–11809

von Ballmoos C, Cook GM, Dimroth P (2008) Unique rotary ATP synthase and its biological diversity. Annu Rev Biophys 37: 43–64

von Heijne G (2006) Membrane-protein topology. Nat Rev Mol Cell Biol 7: 909–918

Wächtershäuser G (2003) From pre-cells to Eukarya – a tale of two lipids. Mol Microbiol 47: 13–22

Walker JE (1998) ATP synthesis by rotary catalysis. Angew Chem 37: 2309–2319

Weber J and Senior AE (2003) ATP synthesis driven by proton transport in F_1F_0-ATP synthase. FEBS Lett 545: 61–70

White SH and von Heijne G (2008) How translocons select transmembrane helices. Annu Rev Biophys 37: 23–42

Wieczorek H, Brown D, Grinstein S, Ehrenfeld J, Harvey WR (1999) Animal plasma membrane energization by proton-motive V-ATPases. Bioessays 21: 637–648

Williams RJP and Frausto da Silva JJR (2006) The chemistry of evolution: the development of our ecosystem. Elsevier, Amsterdam

Woese C (1998) The universal ancestor. Proc Natl Acad Sci USA 95: 6854–6859

Woese CR (1987) Bacterial evolution. Microbiol Rev 51: 221–271

Yanagida S, Kizumoto H, Ishimaru Y, Pac C, Sakurai H (1985) Zinc sulfide catalyzed photochemical conversion of primary amines to secondary amines. Chemistry Lett 14: 141–144

Yokoyama K and Imamura H (2005) Rotation, structure, and classification of prokaryotic V-ATPase. J Bioenerg Biomembr 37: 405–410

Zerkle AL, House CH, Brantley SL (2005) Biogeochemical signatures through time as inferred from whole microbial genomes. Am J Sci 305: 467–502

Zhang XV, Ellery SP, Friend CM, Holland HD, Michel FM, Schoonen MAA, Martin ST (2007) Photodriven reduction and oxidation reactions on colloidal semiconductor particles: implications for prebiotic synthesis. J Photochem Photobiol A Chem 185: 301–311

Zhang Y and Fillingame RH (1995) Changing the ion binding specificity of the *Escherichia coli* H^+-transporting ATP synthase by directed mutagenesis of subunit c. J Biol Chem 270: 87–93

Molecular archeological studies of transmembrane transport systems

Milton H. Saier Jr, Bin Wang, Eric I. Sun, Madeleine Matias and Ming Ren Yen

Division of Biological Sciences, University of California at San Diego, La Jolla, CA, USA

Abstract

We here review studies concerned with the evolutionary pathways taken for the appearance of complex transport systems. The transmembrane protein constituents of these systems generally arose by (1) intragenic duplications, (2) gene fusions, and (3) the superimposition of enzymes onto carriers. In a few instances, we have documented examples of "reverse" or "retrograde" evolution where complex carriers have apparently lost parts of their polypeptide chains to give rise to simpler channels. Some functional superfamilies of transporters that are energized by adenosine triphosphate (ATP) or phosphoenolpyruvate (PEP) include several independently evolving permease families. The ubiquitous ATP-binding cassette (ABC) superfamily couples transport to ATP hydrolysis where the ATPases are superimposed on at least three distinct, independently evolving families of permeases. The prokaryotic sugar transporting phosphotransferase system (PTS) uses homologous PEP-dependent general energy-coupling phosphoryl transfer enzymes superimposed on at least three independently arising families of permeases to give rise to complex group translocators that modify their sugar substrates during transport, releasing cytoplasmic sugar phosphates. We suggest that simple carriers evolved independently of the energizing enzymes, and that chemical energization of transport resulted from the physical and functional coupling of the enzymes to the carriers.

1 Introduction

"Nothing in biology makes sense except in the light of evolution." This precept, enunciated by Dobzhansky (1964), summarizes what must be considered to be one of the most important tenants of biology. We owe our gratitude to the greatest biologist of all times, Charles Darwin, for revealing the essence and implications of this

Corresponding author: Milton H. Saier Jr, Division of Biological Sciences, University of California at San Diego, La Jolla, CA 92093-0116, USA (E-mail: msaier@ucsd.edu)

29

statement. In fact, the scientific method and the Unity Principle in Biology, which allow extrapolation of information from one organism to others, depend on the discoveries of Darwin (Ayala 2009).

Evolution is the framework upon which we must interpret and understand all biological data, but the genomics revolution has tremendously expanded the available dataset. The reams of data are now so large and numerous that we need refined tools to read the many pages in the book of life. The recently founded disciplines of bioinformatics and biosystematics, ever expanding, provide these tools (Koonin 2009).

2 Molecular transport

Focusing on transmembrane molecular transport, we recognize that transport systems are essential to every living cell (Saier 2000). They (1) allow all essential nutrients into the cell and its compartments, (2) regulate concentrations of metabolites by both uptake and excretion mechanisms, (3) provide ion concentration gradients that generate electrical potentials and allow the propagation of action potentials, (4) export macromolecules such as complex carbohydrates, proteins, lipids, DNA, and RNA, (5) catalyze export and uptake of signaling molecules that mediate intercellular and intracellular communication, (6) prevent toxic effects of poisons by catalyzing their active efflux, and (7) participate in biological warfare by exporting biologically active agents that damage or kill other cells. Thus, transport is an essential aspect of all life-endowing processes: metabolism, communication, biosynthesis, reproduction, and both cooperative and antagonistic interorganismal behaviors (Busch and Saier 2002). The Transporter Classification DataBase (TCDB; *www.tcdb.org*) classifies all of the transporters found in nature and presents brief descriptions of their known characteristics (Saier et al. 2006, 2009).

3 Techniques to establish homology or the lack of homology

To establish that different membrane proteins are polyphyletic (arose independently of one another), it is necessary to establish distinct routes of evolutionary appearance (Saier 1994). This has become possible due to the availability of (1) more sensitive software (Yen et al. 2009), (2) large numbers of homologs resulting from genome sequencing, and (3) application of the Superfamily Principle. This principle, first established by Doolittle (1981), states that if A is homologous (derived from a common ancestor) to B, and B is homologous to C, then A must be homologous to C. In spite of the simplicity of this fairly obvious precept, its validity is still questioned by some molecular biologists.

To establish homology between two or more proteins, one can invoke the Superfamily Principle and various statistical approaches (Wang et al. 2009; Yen et al. 2010). Our criteria require that two sequences of at least 60 amino acyl residues give a comparison score of at least 10.0 standard deviations, corresponding to a probability of 10^{-24} that this degree of similarity arose by chance (Dayhoff et al. 1983; Saier 1994). Programs are used that take account of unusual amino acid compositions in these proteins, such as convergently evolved short sequence motifs, multiple short repeat sequences and regions of strong hydrophobicity for membrane proteins (Yen et al. 2009, 2010).

To establish a *lack* of homology, one must show that the two proteins evolved independently, following different evolutionary pathways (Wang et al. 2009). This has recently been achieved for the adenosine triphosphate-binding cassette (ABC) and phosphotransferase system (PTS) functional superfamilies, revealing at least three distinct transmembrane families of transport proteins in each of these superfamilies (Saier et al. 2005; Wang et al. 2009).

4 Transport protein diversity

Our bioinformatics laboratory has conducted bioinformatic analyses of integral membrane transport proteins belonging to dozens of families (Saier 2003a). These families rarely include proteins that function in a capacity other than transport. Only a few of the members of these families can function in other capacities. These other derived functions include regulation and signal transduction (Stasyk et al. 2008; Aguena and Spira 2009; Tetsch and Jung 2009).

We have presented evidence that transporters evolved independently of other protein classes such as enzymes, structural proteins, and regulatory proteins (Saier 2003b). Many transporters have arisen by intragenic duplication, triplication, and quadruplication events, in which the numbers of transmembrane α-helical hydrophobic segments (TMSs) or amphipathic β-strands have increased. The elements multiplied may encode 2, 3, 4, 5, 6, 10, or 12 TMSs and gave rise to proteins with 4, 6, 7, 8, 9, 10, 11, 12, 16, 20, 24, and 30 TMSs. Gene fusion, splicing, deletion, and insertion events have also contributed to protein topological diversity.

Amino acid substitutions have allowed membrane-embedded domains to become hydrophilic domains and vice versa, although this has occurred rarely. Some evidence suggests that amino acid substitutions occurring over evolutionary time may, in some cases, have drastically altered protein structure. Other proteins such as toxins that are inserted into the membranes of target cells, and amyloid proteins that can form transmembrane pores in one conformation, can exist in more than one stable form (Kelly 1998; Rossjohn et al. 2007). These observations reveal that the lack of three-dimensional structural similarities cannot be used as a reliable indication of independent origin.

The results summarized in this chapter establish the independent origins of several transporter families and allow reliable postulation of the specific pathways taken for their appearance. We also present an example of "reverse" evolution whereby a complex transport carrier has lost part of its structure through evolution to give rise to a functionally more simple channel protein. These novel examples provide insight into the diverse mechanisms by which transport systems evolved.

5 The ABC superfamily

The ABC superfamily is considered to be one of the two largest superfamilies of transmembrane transporters found in nature (Davidson et al. 2008; Higgins 2007), the other being the Major Facilitator Superfamily (MFS) of secondary carriers (Pao et al. 1998;

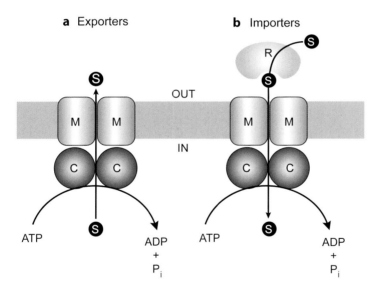

Fig. 1. Generalized schematic depiction of ABC transporters. (a) ABC solute exporters consist of the membrane-embedded porter (M) and the superimposed ATP hydrolyzing subunit on the cytoplasmic side of the membrane (C) that provides energy for efflux of the solute from the cell. These two components can be fused together in a single polypeptide chain to form a half permease. In some ABC porters, especially but not exclusively in eukaryotes, two M and two C equivalent domains are fused together to comprise a full length ABC transport system in one polypeptide chain. (b) ABC solute uptake porters generally consist of the membrane-embedded porter (M), the ABC energizer (C), and an extracytoplasmic receptor (R). R transfers the solute to the porter, M and triggers ATP hydrolysis by C. Phylogenetic analyses of the homologous C subunits have shown that they segregate according to polarity (direction) of transport, and that the ABC proteins that function with exporters, segregate according to the topological type of membrane subunits (ABC1, 2, or 3) they energize.

Saier et al. 1999). The generalized structures of ABC exporters (a) and importers (b) are depicted schematically in Fig. 1 (Davidson et al. 2008; Oldham et al. 2008). By tradition, the ABC superfamily is defined on the basis of its energy-coupling proteins, the monophyletic ATP hydrolyzing (ABC) domains or subunits, which all share a single common origin (the C subunits in Fig. 1; Higgins 2007). The integral membrane porter domains or subunits (the M subunits in Fig. 1) provide the basis for classification of most other transport protein families. In many, but not all ABC uptake transporters, additional components, the extracytoplasmic receptors (the R subunits in Fig. 1b) are essential for efficient activity.

We have long wondered if the membrane constituents of ABC transporters are monophyletic. If they are homologous, a single transport mechanism and common structural features can be predicted. However, if ABC porters are polyphyletic, there is no basis for extrapolating findings made with one phylogenetic group of proteins to another (Chang et al. 2004). With this in mind, we recently investigated the evolutionary origins of ABC exporters (Wang et al. 2009).

6 Independent origins for ABC porters

There are currently 86 recognized families of ABC transporters, 33 for solute uptake and 53 for solute export (see *www.tcdb.org*). They can transport small molecules such as nutrients, salts, and toxins, or they can function in macromolecular efflux, secreting proteins, complex carbohydrates, and lipids (Wang et al. 2009). The integral membrane components (the M subunits in Fig. 1) belong to three topological types called ABC1, ABC2, and ABC3. All three types are present in all three domains of living organisms, bacteria, archaea, and eukaryotes, although one of them (ABC3) has only been found in lower unicellular eukaryotes but not in higher multicellular eukaryotes (Wang et al. 2009).

The three types of porters are shown schematically in Fig. 2. The three repeats of two TMSs, found in the 6-TMS ABC1 porters, arose from a primordial 2-TMS hairpin-encoding genetic element by intragenic triplication, yielding a protein consisting of three tandem transmembrane hairpins, all in the same polypeptide chain, and all with the same orientation in the membrane (Fig. 2a). By contrast, the ABC2 porters arose from a primordial 3-TMS-encoding genetic element by intragenic duplication, yielding the present day 6-TMS proteins with the two homologous halves having opposite orientation in the membrane (Fig. 2b). Finally, ABC3 porters can have 4, 8, or 10 putative TMSs (Khwaja et al. 2005). The 4-TMS-encoding genetic element, present as a pair in some ABC3 transporters, intrageically duplicated to yield the 8- and 10-TMS proteins, always with the two homologous 4-TMS domains having the same orientation in the membrane (Fig. 2c). Surprisingly, in the

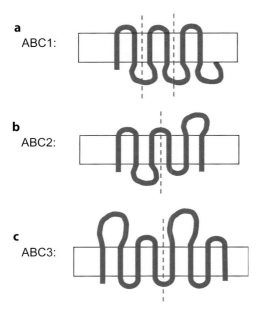

Fig. 2. **Three topological types of ABC porters, illustrating the types of internal repeats present in each one.** In all cases, vertical dashed lines separate the repeat units. (**a**) ABC1: a 6-TMS topology resulting from intragenic triplication of a primordial 2-TMS-encoding genetic element. The three hairpin repeats have the same orientation in the membrane and are homologous to one another. (**b**) ABC2: a 6-TMS topology resulting from intragenic duplication of a primordial 3-TMS-encoding genetic element. The two 3-TMS repeats have opposite orientation in the membrane. (**c**) ABC3: an 8-TMS topology resulting from intragenic duplication of a primordial 4-TMS-encoding genetic element. The two 4-TMS repeats have the same orientation in the membrane. Two 4-TMS half permeases can also comprise an ABC3 porter, similar to ABC1 and ABC2 porters as shown schematically in Fig. 1.

10-TMS proteins, the two repeat units are separated by two extra, non-homologous TMSs that must have arisen during or after the duplication event (Khwaja et al. 2005).

Although the ABC1 porter type is found in larger numbers of exporters than ABC2 types, it appears that all ABC uptake systems are of the ABC2 type. We postulate that the ABC3 porters arose relatively late in evolutionary time, as the repeat units in the 8- and 10-TMS proteins have a high degree of sequence identity compared to the smaller repeat units in the ABC1 and ABC2 types. Moreover, the distribution of ABC3 proteins is much more restricted than those of the other two topological types (see Table 1 in Wang et al. 2009, for more detailed classification, topological assignments, and properties of these primary active transporters).

High-resolution X-ray structures are now available for several ABC importers and several ABC exporters (Davidson et al. 2008; Oldham et al. 2008). The original research references describing these structures can be found in these cited references as well as in Wang et al. (2009). The importers all have a similar three dimensional fold, but this structure differs drastically from that of the exporters, which however show similar three dimensional structures relative to each other. This observation is in agreement with our conclusion of multiple origins for ABC1-3 porters. This is because all of the importers prove to be of the ABC2 type, while all of the exporters for which structural information is currently available are of the ABC1 type. Structures of ABC2 and ABC3 exporters are not yet available. X-ray crystallographers therefore have their work cut out for them. We are betting that the structures of these porters will confirm our bioinformatic conclusions regarding their independent origins.

7 The phosphoenolpyruvate-dependent sugar transporting phosphotransferase system (PTS)

PTS transporters are structurally even more complex than ABC transporters (Fig. 3). This fact may be attributed to the functional complexity of the system, as it

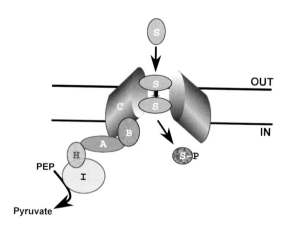

Fig. 3. Schematic depiction of the protein constituents of a typical PTS permease. A PTS permease is a sugar transporting Enzyme II complex of the bacterial phosphoenolpyruvate-dependent phosphotransferase system. The sugar substrate (S) is transported from the extracellular medium through the membrane in a pathway determined by the integral membrane permease-like Enzyme IIC (C) constituent, often a homodimer in the membrane as shown. The sequentially acting energy-coupling proteins transfer a phosphoryl group from the initial phosphoryl donor, phosphoenolpyruvate (PEP), to the ultimate phosphoryl acceptor, extracellular sugar, yielding intracellular sugar–phosphate (S–P). These enzymes are: Enzyme I (I), HPr (H), Enzyme IIA (A), and Enzyme IIB (B). I, first general energy-coupling protein; H, second general energy-coupling protein; A, indirect family-specific phosphoryl donor; B, direct permease-specific phosphoryl donor; and C, permease that catalyzes transport and phosphorylation of the sugar substrate. A given bacterial cell may possess multiple PTS Enzyme II complexes, each specific for a different set of sugars. Only the Mannose-type systems have the IID components.

Table 1. The PTS: functional complexity as indicated by the processes it catalyzes

1.	Chemoreception
2.	Transport
3.	Sugar phosphorylation
4.	Protein phosphorylation
5.	Regulation of non-PTS transport
6.	Regulation of carbon metabolism
7.	Coordination of nitrogen and carbon metabolism
8.	Regulation of gene expression
9.	Regulation of pathogenesis
10.	Regulation of cell physiology

has at least ten recognized functions (see Table 1 and Barabote and Saier 2005). It functions in transport, sugar phosphorylation, chemoreception, and the regulation of both gene expression and metabolism by several mechanisms. Regulation is me-

Table 2. Characteristics of three families of PTS permease complexes[a]

A. The Glc–Fru–Lac family
1. Fru: The original PTS (proposed)
2. Mosaic origins of IIAs and IIBs
3. IIA^{Glc} is not homologous to IIA^{Mtl}
4. IIB^{Glc} is not homologous to IIB^{Chb}
B. The Asc–Gat family
1. IIC^{Asc} homologs are often fused to IIA and IIB homologs, but IIC^{Gat} homologs never are
2. IIC^{Asc} homologs are always encoded by genes in operons with IIA and IIB genes, but IIC^{Gat} homologs can be encoded in operons lacking IIA and IIB genes
3. Some IIC^{Gat} homologs are found in organisms that lack all other PTS proteins
4. Asc and Gat IIA and IIB constituents are distantly related to IIA and IIB constituents of the Glc–Fru–Lac family
C. The Man family
1. All constituents (IIA, IIB, IIC, and IID) differ structurally from all other PTS permease proteins
2. All members, but only members of this family, have IID constituents
3. The IIB constituents are phosphorylated on histidine rather than on cysteine as is true for all other IIBs
4. All constituents of the Mannose IIABCD systems probably evolved independently of those of the IIABC systems

[a]The superscripted abbreviations refer to the substrate sugars as follows: Glc, glucose; Fru, fructose; Lac, lactose; Mtl, mannitol; Chb: diacetyl chitobiose; Asc, L-ascorbate; Gat, galactitol; Man, mannose.

diated by the capacity of the PTS to serve as a protein kinase system, phosphorylating numerous regulatory proteins. Additionally, the PTS functions in regulation and is regulated by virtue of its capacity to become phosphorylated by ATP-dependent protein kinases (Deutscher et al. 2006; Saier et al. 2005).

A PTS permease complex generally consists of five proteins or protein domains called Enzyme I, HPr, IIA, IIB, and IIC (Barabote and Saier 2005). As shown in Fig. 3, the ultimate phosphoryl donor is phosphoenolpyruvate (PEP) which phosphorylates Enzyme I, an enzyme showing extensive sequence similarity to an enzyme of gluconeogenesis, PEP synthase. The phosphoryl group of Enzyme I is transferred to HPr, and then it is sequentially passed on to two constituents of the Enzyme II complex, IIA, and IIB (Postma et al. 1993). Only when IIB is phosphorylated can the sugar be transported into the cell, and it is simultaneously phosphorylated at the expense of IIB-phosphate. Surprisingly, X-ray crystallographic data for the Enzymes IIA and IIB have shown that they do not all have the same fold. Instead, it appears that there are three structurally dissimilar IIAs and three structurally dissimilar IIBs. The IIA and IIB constituents may therefore be polyphyletic (Table 2; see Saier et al. 2005; and Peterkofsky et al. 2001).

8 Independent origins for PTS permeases

Evidence suggests that, like ABC-type porters, the IIC components of PTS permeases are polyphyletic (Hvorup et al. 2003; Saier et al. 2005). There are three topologically, and presumably evolutionarily distinct, families of PTS porters. These will be presented briefly below.

Members of the Glucose (Glc)–Fructose (Fru)–Lactose (Lac) family transport a wide range of sugars and have a uniform topology of eight TMSs per polypeptide chain (Nguyen et al. 2006). These porters function with two of the three recognized types of IIA and IIB proteins. These systems were probably the first to evolve. Fructose systems are thought to be primordial because (1) many bacteria have only this PTS permease, (2) only fructose 1- and 6-phosphates are fed directly into glycolysis without modification, and (3) the PTS may have evolved as a component of glycolysis.

The PTS lactose permease is unrelated to the more carefully studied lactose permease (LacY) of *Escherichia coli*, which is a secondary carrier and a member of the MFS (see above and Pao et al. 1998; Saier et al. 1999). Preliminary results suggest that many of the 8-TMS PTS porters have arisen from a 3-TMS precursor by duplication to 6, as did the ABC2 porters, but then 2 TMSs were added at the C-terminus (unpublished data). The details of this proposed evolutionary pathway are still under investigation.

The ascorbate (Asc)–galactitol (Gat) family of IIC proteins consists of members with 12 TMSs showing no sequence similarity to members of the Glc–Fru–Lac family (Hvorup et al. 2003). The properties summarized in Table 2 suggest that the Gat porters can function either as secondary carriers or as PTS-coupled systems, while Asc porters can function only as PTS-coupled systems (Saier et al. 2005). These 12-TMS IIC permeases arose by internal duplication of a genetic element encoding a primordial 6-TMS-encoding element. These permeases function with sequence divergent IIA and IIB constituents derived from those of the Fru–Glc–Lac family (Saier et al. 2005).

Finally, the primary membrane constituents of the third family, the Mannose (Man) family, consist of 6-TMS proteins showing no significant sequence similarity to proteins of the other two PTS porter families (Huber and Erni 1996). Additionally, they function with another integral membrane protein called IID. IID proteins have just one TMS, and although they are essential for function of the mannose-type PTS permeases, they are lacking in the other types of PTS permeases (Huber and Erni 1996; Esquinas-Rychen and Erni 2001). Moreover, their IIA and IIB constituents (often fused together) are also unrelated to those of the other two families. It seems that *all* constituents of the Mannose PTS transporters evolved completely independently of those of the other two families of PTS porters. Their only shared components are the general energy-coupling proteins, Enzyme I and HPr (Postma et al. 1993).

It should be noted, that the basic unit in all three types of porters contains 6-TMSs as described above, and the routes of appearance of these units are not yet established. They may conceivably have had an early common origin. If so, they diverged into three different topological types during the evolutionary process. X-ray crystallographic studies as well as further bioinformatic analyses may shed light on this interesting possibility.

9 Reverse (retro)-evolution

As noted in the Introduction Section, "evolution tends towards complexity". Recently, a novel family of Ca^{2+} release-activated Ca^{2+} (CRAC) channels, promoting an immune response to pathogens, has been characterized (Matias et al. 2010; Parekh 2009). The proteins comprising these channels are called Orai. Each such polypeptide chain possesses four transmembrane helical segments (TMSs; Matias et al. 2010), and these assemble into an oligomeric channel complex (Di Capite and Parekh 2009; Prakriya 2009).

We have provided convincing evidence that the 4-TMS Orai channels are homologous to parts of the members of a family of heavy metal ion (Me^{2+}) exporters

Table 3. Comparisons of CDF carriers with crac channels

	CDF (TC# 2.A.4)	Crac-C (TC# 1.A.52)
1.	Secondary carriers: catalyze Me^{2+}:H$^+$ antiport	Channels: catalyze bidirectional Ca^{2+} flux
2.	Ubiquitous; in plasma membrane and intracellular membranes of eukaryotes	Present only in eukaryotes; at plasma membrane/endoplasmic reticulum junctions
3.	6-TMSs; N- and C-termini inside; dimeric	4-TMSs; N- and C-termini inside; tetrameric
4.	Much size and sequence divergence	Little size and sequence divergence
5.	Two aspartates are critical for Me^{2+} binding	Two glutamates are critical for Ca^{2+} binding

(Matias et al. 2010). This family of carriers, the Cation Diffusion Facilitator (CDF) family, exhibits the properties summarized in Table 3, which also compares these properties with those of the CRAC channel family: (1) while CDF carriers catalyze Me^{2+}:H$^+$ exchange (antiport), CRAC channels catalyze bidirectional Ca^{2+} flux; (2) while CDF carriers are ubiquitous, being present in bacteria, archaea, and eukaryotes, CRAC channels are present only in eukaryotes and are found specifically at the plasma membrane/endoplasmic reticulum junction; (3) while CDF carriers have 6-TMSs, CRAC channels have just four per polypeptide chain, both with N- and C-termini in the cytoplasm; (4) while CDF carriers exhibit tremendous size and sequence divergence, CRAC channels are relatively of the same size and exhibit a much higher degree of sequence similarity; and (5) finally, both use a pair of anionic amino acid residues (aspartate or glutamate) to bind the divalent cation. All of these properties indirectly suggest some type of evolutionary relationship between these two protein types.

Careful examination of the sequences of CDF carriers and Orai CRAC channels revealed highly significant sequence similarity (Matias et al. 2010). While TMSs 3 and 4 of CDF proteins show statistically significant similarity with TMSs 1 and 2 of the Orai channels, and the same is true of TMSs 5 and 6 of CDF carriers versus TMSs 3 and 4 of Orai proteins, other combinations of comparisons showed much lower comparison scores. It could be shown that both protein types derived from a primordial 2-TMS hairpin structure, triplicating (via intragenic triplication) to give 6-TMSs for the CDF carriers and apparently duplicating (via intragenic duplication) to give 4-TMS proteins for the Orai CRAC channels. These facts as well as the ubiquitous distribution and the tremendous size and sequence divergence of the former proteins versus the limited distribution of the much more uniform Orai proteins, which exhibit much more similar sizes and sequences, led to the proposal shown in Fig. 4.

Figure 4 shows the proposed pathway for the origin of the 6-TMS CDF proteins and for the CDF-derived 4-TMS Orai proteins. Intragenic triplication of a primor-

39

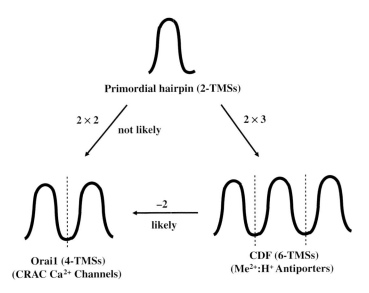

Primordial hairpin (2-TMSs)

2×2 / not likely

2×3

−2
likely

Orai1 (4-TMSs)
(CRAC Ca^{2+} Channels)

CDF (6-TMSs)
(Me^{2+}:H$^+$ Antiporters)

Fig. 4. Proposed common origin for CRAC channels and CDF carriers. The figure illustrates two potential pathways: the likely pathway whereby triplication of the primordial hairpin structure gave rise to a 6-TMS CDF carrier, followed by the loss of TMSs 1–2 to give 4-TMS Orai channels. See text for details.

dial 2-TMS-encoding element yielded the 6-TMS carriers, and subsequent loss of the gene segment coding for the first two of these 6 TMSs resulted in the fundamentally less complex channel proteins of the CRAC family (Matias et al. 2010). This seems to be an example of "reverse" or "retrograde" evolution where a more complex structure yields a smaller and fundamentally simpler one. This is in contrast to the usual pathway of evolution where smaller, simpler structures, via intragenic duplication and fusion events, give rise to structurally and functionally more complex proteins (Saier 2003b). This is the first example we know of where a complex carrier was probably the precursor of a simpler channel-forming protein, although other types of simplifying mutations have been described.

10 Conclusions and perspectives

Recent quantitative bioinformatic data have revealed that two of the largest functional superfamilies of transporters, ABC-type primary active transporters and PTS-type substrate phosphorylating transporters, are true mosaic systems. Both apparently consist of at least three families of permease subunits or domains that have evolved independently of one another, following different routes of evolutionary appearance. Their sole common feature is their use of sequence-similar types of energy-coupling proteins, the ATP hydrolyzing subunits of ABC transporters, and the common phosphoryl transfer proteins, Enzyme I and HPr, of PTS porters. It is interesting to note, however, that the system-specific IIA and IIB constituents, like the IIC permeases, may have arisen at least three times independently. Moreover, the PTS can also phosphorylate dihydroxyacetone using a soluble, cytoplasmic Enzyme II complex that, surprisingly, is derived from ATP-dependent dihydroxyacetone kinases (Gutknecht et al. 2001).

New evidence is revealing that in both the ABC and PTS superfamilies, the membrane porters alone may be able to catalyze transport without energy coupling involving ATP or PEP (Saier et al. 2005; Eudes et al. 2008; Rodionov et al. 2009; Zhang et al. 2009). Thus, the primordial transport proteins, upon which a chemical form of energy has been superimposed, may still exist in nature, in states resembling their primordial chemical energy uncoupled states. Further studies will be required to determine to what degree these observations are applicable to other complex transport systems. Nevertheless, the bioinformatic analyses reviewed here provide guides for future structural, mechanistic, and physiological investigations.

References

Aguena M and Spira B (2009) Transcriptional processing of the pst operon of *Escherichia coli*. Curr Microbiol 58: 264–267

Ayala FJ (2009) Darwin and the scientific method. Proc Natl Acad Sci USA 106 (Suppl 1): 10033–10039

Barabote RD and Saier MH Jr (2005) Comparative genomic analyses of the bacterial phosphotransferase system. Microbiol Mol Biol Rev 69: 608–634

Busch W and Saier MH Jr (2002) The transporter classification (TC) system, 2002. Crit Rev Biochem Mol Biol 37: 287–337

Chang AB, Lin R, Keith Studley W, Tran CV, Saier MH Jr (2004) Phylogeny as a guide to structure and function of membrane transport proteins. Mol Membr Biol 21: 171–181

Davidson AL, Dassa E, Orelle C, Chen J (2008) Structure, function, and evolution of bacterial ATP-binding cassette systems. Microbiol Mol Biol Rev 72: 317–364 (Table of contents)

Dayhoff MO, Barker WC, Hunt LT (1983) Establishing homologies in protein sequences. Meth Enzymol 91: 524–545

Deutscher J, Francke C, Postma PW (2006) How phosphotransferase system-related protein phosphorylation regulates carbohydrate metabolism in bacteria. Microbiol Mol Biol Rev 70: 939–1031

Di Capite J and Parekh AB (2009) CRAC channels and Ca^{2+} signaling in mast cells. Immunol Rev 231: 45–58

Doolittle RF (1981) Similar amino acid sequences: chance or common ancestry? Science 214: 149–159

Esquinas-Rychen M and Erni B (2001) Facilitation of bacteriophage lambda DNA injection by inner membrane proteins of the bacterial phosphoenol-pyruvate: carbohydrate phosphotransferase system (PTS). J Mol Microbiol Biotechnol 3: 361–370

Eudes A, Erkens GB, Slotboom DJ, Rodionov DA, Naponelli V, Hanson AD (2008) Identification of genes encoding the folate- and thiamine-binding membrane proteins in Firmicutes. J Bacteriol 190: 7591–7594

Gutknecht R, Beutler R, Garcia-Alles LF, Baumann U, Erni B (2001) The dihydroxyacetone kinase of *Escherichia coli* utilizes a phosphoprotein instead of ATP as phosphoryl donor. EMBO J 20: 2480–2486

Higgins CF (2007) Multiple molecular mechanisms for multidrug resistance transporters. Nature 446: 749–757

Huber F and Erni B (1996) Membrane topology of the mannose transporter of *Escherichia coli* K12. Eur J Biochem 239: 810–817

Hvorup R, Chang AB, Saier MH Jr (2003) Bioinformatic analyses of the bacterial L-ascorbate phosphotransferase system permease family. J Mol Microbiol Biotechnol 6: 191–205

Kelly JW (1998) The alternative conformations of amyloidogenic proteins and their multi-step assembly pathways. Curr Opin Struct Biol 8: 101–106

Khwaja M, Ma Q, Saier MH Jr (2005) Topological analysis of integral membrane constituents of prokaryotic ABC efflux systems. Res Microbiol 156: 270–277

Koonin EV (2009) Evolution of genome architecture. Int J Biochem Cell Biol 41: 298–306

Matias M, Gomolplitant KM, Tamang DG, Saier MH (2010) Animal Ca^{2+} release-activated Ca^{2+} (CRAC) channels are homologous to and derived from the ubiquitous cation diffusion facilitators. BMC Research Notes (in press)

Nguyen TX, Yen MR, Barabote RD, Saier MH Jr (2006) Topological predictions for integral membrane permeases of the phosphoenolpyruvate: sugar phosphotransferase system. J Mol Microbiol Biotechnol 11: 345–360

Oldham ML, Davidson AL, Chen J (2008) Structural insights into ABC transporter mechanism. Curr Opin Struct Biol 18: 726–733

Pao SS, Paulsen IT, Saier MH Jr (1998) Major facilitator superfamily. Microbiol Mol Biol Rev 62: 1–34

Parekh AB (2009) Local Ca^{2+} influx through CRAC channels activates temporally and spatially distinct cellular responses. Acta Physiol (Oxf) 195: 29–35

Peterkofsky A, Wang G, Garrett DS, Lee BR, Seok YJ, Clore GM (2001) Three-dimensional structures of protein–protein complexes in the *E. coli* PTS. J Mol Microbiol Biotechnol 3: 347–354

Postma PW, Lengeler JW, Jacobson GR (1993) Phosphoenolpyruvate:carbohydrate phosphotransferase systems of bacteria. Microbiol Rev 57: 543–594

Prakriya M (2009) The molecular physiology of CRAC channels. Immunol Rev 231: 88–98

Rodionov DA, Hebbeln P, Eudes A, ter Beek J, Rodionova IA, Erkens GB, Slotboom DJ, Gelfand MS, Osterman AL, Hanson AD, Eitinger T (2009) A novel class of modular transporters for vitamins in prokaryotes. J Bacteriol 191: 42–51

Rossjohn J, Polekhina G, Feil SC, Morton CJ, Tweten RK, Parker MW (2007) Structures of perfringolysin O suggest a pathway for activation of cholesterol-dependent cytolysins. J Mol Biol 367: 1227–1236

Saier MH Jr (1994) Computer-aided analyses of transport protein sequences: gleaning evidence concerning function, structure, biogenesis, and evolution. Microbiol Rev 58: 71–93

Saier MH Jr (2000) A functional–phylogenetic classification system for transmembrane solute transporters. Microbiol Mol Biol Rev 64: 354–411

Saier MH Jr (2003a) Answering fundamental questions in biology with bioinformatics. ASM News 69: 175–181

from PDB by using the knowledge of experts, who select TMP by hand or to develop new computer algorithms to do this task automatically. We will discuss the manually collected data sets and the developed algorithms in the future sections.

2.2 Manually curated structure resources of TMPs

Manually curated structure resources of TMPs are listed in Table 1. The first such collection was made by Preusch et al. (1998), comprising 24 structures and was later revisited by White and Wimley (1999). The resulted list of PDB files forms the base of the most accurate and constantly maintained Internet resource of membrane proteins of known 3D structures, the so-called White's Database (*http://blanco.biomol. uci.edu/Membrane_Proteins_xtal.html*). In 2006, Hartmut Michel collected a list of membrane proteins from PDB but this list has not been maintained so far, therefore the information in it is rather out of date (*http://www.mpibp-frankfurt.mpg.de/ michel/public/memprotstruct.html*).

The Membrane Protein Data Base (MPDB) was created by manually curating all PDB entries having "membrane" keywords (Raman et al. 2006). Information, such as protein characteristics, structure determination method, crystallization technique, detergent, temperature, pH, author, etc. have been extracted from the PDB headers into a relational database, which in turn can be easily used for various searches and statistics. Record entries are hyperlinked to the PDB and Pfam for viewing sequence, 3D structure and domain architecture, and for downloading coordinates. Links to PubMed are also provided. It was an intention to update the database weekly, following the PDB updates, but in the last years, the database was maintained irregularly.

Scientists can use the various structure classification databases for selecting and investigating TMPs as well. Such database is SCOP (Lo Conte et al. 2000), in which the "Membrane and cell surface proteins and peptides" class contains the vast majority of TMPs whose structures have been determined so far. The other widely used classification database, the CATH database (Orengo et al. 1998) does not contain

Table 1. Manually curated structure resources of TMPs

Name and URL	Last update[a]	No. of entries
Hartmut Michel's Database (*http://www.mpibp-frankfurt. mpg.de/michel/public/memprotstruct.html*)	March 30, 2006	93
MPDB (*http://www.mpdb.ul.ie*)	September 8, 2009	1005
MPtopo (*http://blanco.biomol.uci.edu/mptopo*)	August 30, 2007	25
Stephen White's Database (*http://blanco.biomol.uci. edu/Membrane_Proteins_xtal.html*)	October 29, 2009	559

[a]Before December 10, 2009.

47

such class definition; therefore users can apply only for keyword searching to select membrane proteins in this classification system. While SCOP is updated regularly, CATH update has been done over 1 year ago.

2.3 TMDET algorithm

The above-mentioned difficulties of PDB and the fact that the manually created Internet sites and databases comprise list of TMPs, but do not give information about the membrane embedded part of these proteins, necessitate to develop an automatic algorithm, called TMDET algorithm (Tusnády et al. 2004). TMDET is a geometrical approach, which is able to distinguish between transmembrane and globular proteins using structural information only and can locate the most likely position of the lipid bilayer, i.e., the membrane planes relative to the position of atomic coordinates. This information is absent in the PDB files, because during the structure determination of TMPs one vital component, the membrane itself is missing from these structures, as TMPs are taken out from the lipid bilayer, and crystalized by masking their exposed hydrophobic parts by amphiphilic detergents, so that the protein–detergent complex can be treated similarly to soluble proteins (Ostermeier and Michel 1997). The discrimination work even in cases of low resolution or incomplete structures such as fragments or parts of large multi-chain complexes.

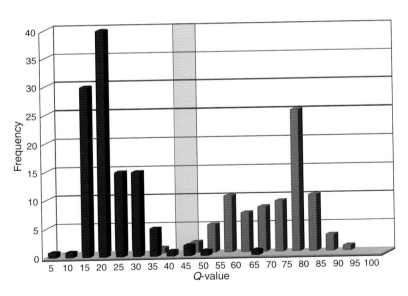

Fig. 1. Discrimination between globular (blue bars) and transmembrane (red bars) proteins using the objective value (Q-value) of TMDET algorithm.

The algorithm utilizes the basic properties of TMPs that is the membrane embedded part of a TMP contains regular secondary structures (α-helices or β-strands) and the protein surface exposed to the lipids is hydrophobic. To find, which part of a TMPs are in the double lipid layer, an objective function was defined and optimization was made to maximize its value by rotating the TMP under investigation through the 4π direction of the space. In a given direction, the protein is cut into 1 Å

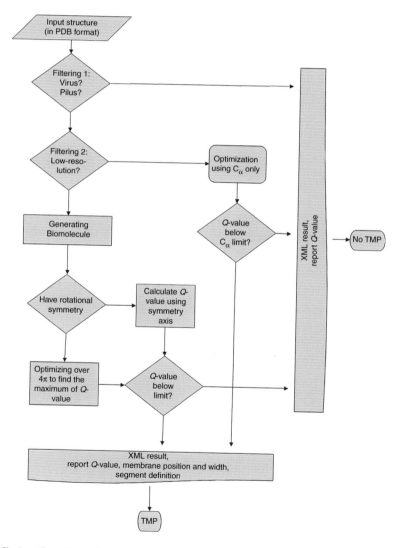

Fig. 2. Flow chart of TMDET algorithm.

wide parallel slices, and the relative frequencies of secondary structure elements and the membrane-exposed "water accessible surface area" are calculated in each slice. The weighted sum of these values is averaged through the slices, and the direction and position giving the maximum value of these averages gives the most probable position of the membrane, while the final value of the objective function can be used to make differences between transmembrane and globular proteins, as it can be seen on Fig. 1. The main steps of TMDET algorithm are shown in Fig. 2.

One obvious power of the TMDET algorithm, that it can identify non-biological quaternary structures, by a special symmetry investigation of protein chains. Non-biological oligomer structures can be formed during the crystalization process. For example, in the PDB structure 1f88 of bovine rhodopsin, two identical chains are arranged in parallel, but in a non-biological, head-to-tail orientation (Fig. 3; Palczewski et al. 2000). This arrangement allows strong contacts between the trans-membrane helices as well as the soluble parts of the molecule, but because of the differences of the extracellular and intracellular environment, this orientation cannot be the native complex. Nevertheless, the significant number of interactions makes it difficult to distinguish these cases from real oligomeric structures. The server for the quaternary structure of proteins originally developed for globular proteins (PQS; Henrick and Thornton 1998), failed to recognize these artifacts, and this wrong oligomeric structure can be found in the PDB biounit classification as well. However, the TMDET algorithm can recognize the wrong oligomeric form and correctly locate the biological molecule with the correct membrane position.

We have implemented a web server running the TMDET algorithm for user submitting 3D coordinate of TMPs (Tusnády et al. 2005a), which is not in PDB. This can be a newly determined structure or results of modeling. The server accept coor-

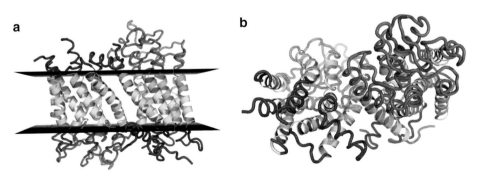

Fig. 3. **The structure of bovine rhodopsin (1f88) from membrane plane (a) and from the extra-cytosolic space (b),** using the following color scheme: membrane-spanning region is colored by yellow, extracellular sides are colored by blue and intracellular sides are colored by red for both chains.

dinates in PDB format and results in an normal HTML output plus a XML file containing the calculated position of membrane planes as well as the necessary rotation and translation so the normal of the membrane plane became parallel to the z-axis. Pictures showing the calculated membrane embedded parts of the protein and the membrane planes are also generated.

2.4 PDBTM database

The TMDET algorithm allows the appropriate classification not only of the well-determined structures, but also of the classification of low-resolution structures, as well as cases when only a fragment or a partial structure of a multi-chain protein complex is available. The discrimination power of TMDET algorithm is >98% (see Fig. 1); therefore, it can be used to create and maintain a database that contains the TMPs of known 3D structures, as well as the calculated most probable membrane localization of these proteins. By scanning all entries in the PDB by the TMDET algorithm we have created the PDBTM database (Tusnády et al. 2004, 2005a,b).

The aim of PDBTM database was twofold. First, it assigns a transmembrane character for each entry in the PDB, which allows the construction of a comprehensive and up-to-date list of TMPs with known structures (and obviously a list of non-TMPs too). Second, it identifies the location of the lipid bilayer that is relative to the

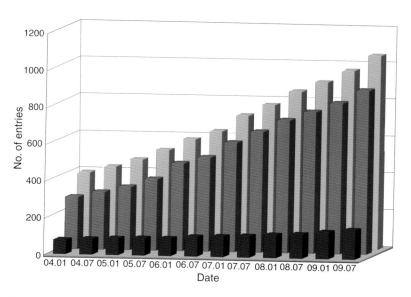

Fig. 4. **Growing statistics of the PDBTM database.** Blue bars: β-barrel proteins; orange bars: α-helical proteins; yellow bars: all transmembrane proteins.

51

coordinates system of the molecule. For all identified TMPs, the algorithm also determines the number and type of transmembrane segments in the sequence as well. Moreover, the algorithm is able to assign canonical and non-canonical elements within the membrane region. Some alpha helical proteins contain segments that do not cross the membrane, but turn back at the middle of the membrane (membrane-loop) (e.g., aquaporin, 1fqy, Murata et al. 2000), whereas the inside of the pore that is formed by beta-barrel proteins are basically shielded from the membrane, and can accommodate globular regions or alpha helical segments as well (e.g., the translocator domain of a bacterial autotransporter, 1uyo, Oomen et al. 2004).

The PDBTM database is available at *http://pdbtm.enzim.hu*. The database content is updated every week following the regular PDB updates by running the TMDET algorithm on each new PDB file. The growing statistics can be seen on Fig. 4. The database is downloadable, either the full database in one big file, or its subsets containing the alpha helical or the beta-barrel TMPs. There is a possibility to browse all the files as well.

2.5 OPM database

While PDBTM focuses on collecting all TMPs from PDB and annotate their sequences, an alternatively developed method (Lomize et al. 2006a) and the resulted Orientation of Proteins in Membranes database (OPM; Lomize et al. 2006b) take more care about the exact position of lipid bilayer in all type of membrane embedded proteins (integral and peripheral membrane proteins). The applied method combines atomic solvation parameters for the water-decadiene system, interfacial polarity profiles in membranes determined in EPR studies, ionization energies of charged residues and elimination of energetic contributions from any atoms situated in the polar pores or channels of TM proteins that do not interact with lipids (Lomize et al. 2006a). This approach can also discriminate between TM and water-soluble proteins and determine the positions of TM proteins with a precision of ~1 Å for the hydrophobic thickness and ~2° for the tilt angle relative to the membrane normal. The results are in good agreement with experimental studies of 24 TMPs. The database provides a classification of membrane proteins as well, based on the structure of their main membrane-associated domains, by using four hierarchical levels: type (TM or peripheral/monotopic protein and peptides), class (all-α, all-β, α + β, α/β), superfamily (evolutionarily related proteins), and family (proteins with clear sequence homology).

2.6 Modeling protein–lipid assembly

Many biophysical studies indicate interactions with lipid/detergent molecules and show that interactions are critical to the folding and stability of membrane proteins.

To better understand the structure function relationship of membrane proteins, the investigation of interactions between protein and lipid environment is needed. We can do it by using molecular dynamics simulations on computers. However, full atomic simulations are extremely time consuming and are limited to system size of up to ~100,000 atoms and time length of up to ~100 ns, therefore, for high-throughput approach it cannot be used. To predict membrane protein interactions with a lipid bilayer a technique with extended size and timescale is needed, such that coarse-grained molecular dynamics (CG-MD) simulations (Bond and Sansom 2006; Sansom et al. 2008). It was shown that there is a good agreement between the atomic simulation and CG-MD simulation in terms of predicted lipid head-group contacts.

CG-MD method was applied on a representative set of 91 proteins embedded in dipalmitoyl-phosphatidylcholin (DPPC) bilayer (Scott et al. 2008). The 91 TMPs included 33 β-barrel and 58 α-helical TMPs. The results of these predictions, i.e., the positioning of the 91 proteins in a membrane bilayer, are collected in a database, called CGDB (*http://sbcb.bioch.ox.ac.uk/cgdb*). For each protein in this database, the coordinates of the lipid and protein at the end of the simulation are available for download, along with the results of analysis of lipid–protein interactions. It was shown that CG approach represents an intermediate resolution, lower than that of all atomic simulations, but provides more details on protein–lipid interactions than is possible with a hydrophobic "slab" models.

3 2D structure resources

TMPs have a common structure, where the transmembrane α-helices are roughly in parallel orientation to each other forming an α-helical bundle in the case of α-helical TMPs, while β-sheets form barrel in the case of β-barrel proteins. Therefore, the knowledge of sequential positions of these transmembrane segments together with the knowledge of what part of the sequence can be found at the in/out side give a rough picture about the structure of these proteins. This information called topology or 2D structures (Elofsson and von Heijne 2007). Therefore, obtaining information about topology helps to understand the structure of TMPs, as well.

Although the various labs make huge efforts to solve more and more structures of membrane proteins, and funded consortiums, like European Membrane Protein Consortium (E-MeP, *http://www.e-mep.org*), New York Consortium on Membrane Protein Structure (NYCOMPS, *http://www.nycomps.org*) or Membrane Protein Structure Initiative (MPSi, *http://www.mpsi.ac.uk*) activated hundreds of trials to express, solubilize, purify, crystalize, and solve structures of membrane proteins, the topology prediction of these proteins is still the easiest, fastest, and complete, but less accurate way to get to know the 2D structures of membrane proteins. There are numerous prediction algorithms and servers to do this task, to overview them we refer to reviews in this area (Chen and Rost 2002; Chen et al. 2002; Elofsson and von Heijne 2007; Punta et al. 2007; Tusnády and Simon 2009). Here, we focus our attention to those databases and algorithms, which collect data about topology and can incorporate these data to make more accurate 2D prediction of TMPs, respectively.

3.1 TOPDB database

There are numerous molecular biological and biophysical methods that can help to learn the topology of TMPs (van Geest and Lolkema 2000). The most often used such methods are fusion experiments [e.g., fusion with alkaline-phosphatase (Boyd et al. 1993), β-lactamase (Broome-Smith et al. 1990), or β-galactosidase (Miller 1972)], topology determination by post-translational modifications (e.g., glycosylation, Bamberg and Sachs 1994), experiments using proteases (Bakos et al. 1996), various techniques using immunolocalization (e.g., epitope insertion, Kast et al. 1996), and experiments utilizing chemical modification techniques (e.g., cystein scanning mutagenesis, Bogdanov et al. 2005). Although there are hundreds of articles using these techniques to determine topologies of TMPs, the collection of these data had not been performed for a long time. There were several database collecting "well characterized" membrane proteins and the topologies of them, such that the so-called Möller database (Möller et al. 2000) and TMPDB database (Ikeda et al. 2003), but those databases do not contain the raw results of the experiments, just the most likely position of the transmembrane segments in the sequence. The authors of these databases underlined that the interpretation of individual experiments are sometimes difficult and the transmembrane annotation was provided by human experts, considering the results of the mere hydropathy plot analysis and experiments. None of these databases were ever updated; therefore, they cannot be used in a recent topology analysis, as the information in them is rather outdated.

The first and almost complete collection of topologic data of TMPs is the TOPDB database (Tusnády et al. 2008). This database contains information gathered from the literature and from public databases available on the Internet for more than a thousand TMPs. The collected raw experimental data are classified and processed uniformly in TOPDB, and this collection and data classification are rather valuable by themselves. A large part of the data come from PDBTM database by translating the 3D arrangement of the lipid bilayer represented in PDBTM into sequential information for TOPDB and by adding information on sidedness (i.e., which part of the 3D structure is outside and which are inside). This information was extracted from literature by checking several hundred articles describing the 3D structure of

TMPs. A large number of structures in the PDB correspond to the soluble fragment of TMPs. These cases also contain information about the topology in an indirect way. The third part of topology data were generated by searching for these solved structures of globular fragments of TMPs in PDB by comparing the corresponding data of a given protein in PDB, PDBTM, and UniProt.

In TOPDB each entry contains the most probable topology of the given protein. The topologies were generated by the modified HMMTOP algorithm (Tusnády and Simon 2001), which is able to use the experimental results as constraints during the topology prediction. We discuss constrained prediction methods later. The web server of the TOPDB database gives the usual access to the database itself, by providing various possibilities to search in topology data (e.g., searching by experiment types, by organism, by keyword, by identifier, or even by structural type), and by giving the opportunity of various downloads of the database. Using the server the user can easily visualize the determined topology of a given protein, as well as the collected experiments for that protein. The user also can find an appropriate and well-documented definition for the various membrane types and which side of each membrane is considered to face to inside and which to outside.

Another more obvious benefit of TOPDB is that we can validate the various experiment types, comparing their results to the 3D structure as the most reliable source of the topology data. In general, the results of various experiments are in good agreement with the 3D structures, but there are two common pitfalls. One of them emerges in the case of utilizing reporter enzymes as fusion proteins. In cases, when the N-terminus is outside, but there was no transmembrane helices prior to the fusion point, the reporter enzyme could not be transferred outside, and remained inactive. Moreover, if the fusion points were after the first transmembrane helix, the topologies were frequently inverted. The other common error is the misinterpretation of the lack of immunoglobulin binding to extra inserted or endogen epitopes localized in the cytosol. In these cases, the binding can be seen only after membrane permeabilization by using special detergents. However, if the extracytosolic epitope is shaded, i.e., the antibody cannot bind due to structural reasons, the epitope will be accessible after the use of detergent, which in turn results in binding.

3.2 TOPDOM database

The TOPDOM database is a collection of domains and sequence motifs located consistently on the same side of the membrane in α-helical TMPs. The database was created by scanning well-annotated TMP sequences in the UniProt database (Bairoch et al. 2005) by specific domain or motif detecting algorithms, such as Prosite (Sigrist et al. 2002), Pfam (Finn et al. 2006), Prints (Attwood et al. 2003), and Smart (Letunic et al. 2004). The identified domains or motifs were added to the database if they were uniformly annotated on the same side of the membrane of the various proteins in the UniProt database.

Because of the continuous and fully automated update in every 3 weeks following the update of UniProt database, the number of identified domains increased from 1010 to 1064, from the launch of the database until now. The database currently contains 463, 155, 302, and 144 domain and motif definitions from Pfam, Prints, Prosite, and Smart databases, respectively, which are uniformly annotated in the same side of the membrane of the various proteins in UniProt database.

The information about the location of the collected domains and motifs can be incorporated into constrained topology prediction algorithms, like HMMTOP, increasing the prediction accuracy, as it will be discussed in Section 3.3.

3.3 Prediction methods incorporating experimental results

Predictions, which can incorporate experimental results or other information about topology, are called constrained predictions. The constrained predictions give results, which correspond to a given condition (i.e., the N-terminal let be at outside) and differs from filtering the output of predictions to a given condition. Constrained prediction can be easily performed using hidden Markov model-based on prediction algorithm by modification of the Baum–Welch and/or Viterbi algorithm. The first such application was HMMTOP2 (Tusnády and Simon 2001). Later two other HMM-based methods, TMHMM and Phobius were also modified to reach this feature (Melén et al. 2003; Käll et al. 2004; Bernsel and von Heijne 2005; Xu et al. 2006). The mathematical details of the necessary modification can be found in Bagos et al. (2006).

Obviously, constrained prediction increases the accuracy and reliability, as it was shown in the case of the human multidrug resistance-associated protein (MRP1; Tusnády and Simon 2001). Later this approach was used to determine the topology of 37 *Saccharomyces cerevisiae* membrane proteins (Kim et al. 2003), global topology analysis of *Escherichia coli* (Daley et al. 2005) and yeast (Kim et al. 2006) genomes and to improve the prediction accuracy by domain assignments (Bernsel and von Heijne 2005). The optimal placement of constraints was also investigated and it was shown that the accuracy can be increased by 10% if the N- or C-terminal of the polypeptide chain is locked, and 20% is the maximum obtainable increase, if one of each loop or tail residue in turn is fixed to its experimentally annotated location (Rapp et al. 2004).

Acknowledgments

Thanks to Monika Fuxreiter for her suggestions and editing the manuscript. This work was supported by grants from Hungarian Research and Development Funds (OTKA) K72569 and K75460. The Charles Simonyi fellowship for I.S. is also gratefully acknowledged.

References

Ahram M, Litou ZI, Fang R, Al-Tawallbeh G (2006) Estimation of membrane proteins in the human proteome. In Silico Biol 6: 379–386

Attwood TK, Bradley P, Flower DR, Gaulton A, Maudling N, Mitchell AL, Moulton G, Nordle A, Paine K, Taylor P, Uddin A, Zygouri C (2003) PRINTS and its automatic supplement, prePRINTS. Nucleic Acids Res 31: 400–402

Bagos PG, Liakopoulos TD, Hamodrakas SJ (2006) Algorithms for incorporating prior topological information in HMMs: application to transmembrane proteins. BMC Bioinform 7: 189

Bairoch A, Apweiler R, Wu CH, Barker WC, Boeckmann B, Ferro S, Gasteiger E, Huang H, Lopez R, Magrane M, Martin MJ, Natale DA, O'Donovan C, Redaschi N, Yeh LL (2005) The Universal Protein Resource (UniProt). Nucleic Acids Res 33: D154–D159

Bakos É, Hegedűs T, Holló Z, Welker E, Tusnády G, Zaman G, Flens M, Váradi A, Sarkadi B (1996) Membrane topology and glycosylation of the human multidrug resistance-associated protein. J Biol Chem 271: 12322–12326

Bamberg K and Sachs G (1994) Topological analysis of H+,K(+)-ATPase using in vitro translation. J Biol Chem 269: 16909–16919

Berman HM, Westbrook J, Feng Z, Gilliland G, Bhat TN, Weissig H, Shindyalov IN, Bourne PE (2000) The protein data bank. Nucleic Acids Res 28: 235–242

Bernsel A and von Heijne G (2005) Improved membrane protein topology prediction by domain assignments. Protein Sci 14: 1723–1728

Bogdanov M, Zhang W, Xie J, Dowhan W (2005) Transmembrane protein topology mapping by the substituted cysteine accessibility method (SCAM(TM)): application to lipid-specific membrane protein topogenesis. Methods 36: 148–171

Bond PJ and Sansom MSP (2006) Insertion and assembly of membrane proteins via simulation. J Am Chem Soc 128: 2697–2704

Boyd D, Traxler B, Beckwith J (1993) Analysis of the topology of a membrane protein by using a minimum number of alkaline phosphatase fusions. J Bacteriol 175: 553–556

Broome-Smith JK, Tadayyon M, Zhang Y (1990) Beta-lactamase as a probe of membrane protein assembly and protein export. Mol Microbiol 4: 1637–1644

Chen CP and Rost B (2002) State-of-the-art in membrane protein prediction. Appl Bioinform 1: 21–35

Chen CP, Kernytsky A, Rost B (2002) Transmembrane helix predictions revisited. Protein Sci 11: 2774–2791

Daley DO, Rapp M, Granseth E, Melén K, Drew D, von Heijne G (2005) Global topology analysis of the *Escherichia coli* inner membrane proteome. Science 308: 1321–1323

Elofsson A and von Heijne G (2007) Membrane protein structure: prediction versus reality. Ann Rev Biochem 76: 125–140

Finn RD, Mistry J, Schuster-Böckler B, Griffiths-Jones S, Hollich V, Lassmann T, Moxon S, Marshall M, Khanna A, Durbin R, Eddy SR, Sonnhammer ELL, Bateman A (2006) Pfam: clans, web tools and services. Nucleic Acids Res 34: D247–D251

van Geest M and Lolkema JS (2000) Membrane topology and insertion of membrane proteins: search for topogenic signals. Microbiol Mol Biol Rev 64: 13–33

Henrick K and Thornton JM (1998) PQS: a protein quaternary structure file server. Trends Biochem Sci 23: 358–361

Henrick K, Feng Z, Bluhm WF, Dimitropoulos D, Doreleijers JF, Dutta S, Flippen-Anderson JL, Ionides J, Kamada C, Krissinel E, Lawson CL, Markley JL, Nakamura H, Newman R, Shimizu Y, Swaminathan J, Velankar S, Ory J, Ulrich EL, Vranken W, Westbrook J, Yamashita R, Yang H, Young J, Yousufuddin M, Berman HM (2008) Remediation of the protein data bank archive. Nucleic Acids Res 36: D426–D433

Ikeda M, Arai M, Okuno T, Shimizu T (2003) TMPDB: a database of experimentally-characterized transmembrane topologies. Nucleic Acids Res 31: 406–409

Jones DT (1998) Do transmembrane protein superfolds exist? FEBS Lett 423: 281–285

Käll L, Krogh A, Sonnhammer ELL (2004) A combined transmembrane topology and signal peptide prediction method. J Mol Biol 338: 1027–1036

Kast C, Canfield V, Levenson R, Gros P (1996) Transmembrane organization of mouse P-glycoprotein determined by epitope insertion and immunofluorescence. J Biol Chem 271: 9240–9248

Kim H, Melén K, von Heijne G (2003) Topology models for 37 *Saccharomyces cerevisiae* membrane proteins based on C-terminal reporter fusions and predictions. J Biol Chem 278: 10208–10213

Kim H, Melén K, Osterberg M, von Heijne G (2006) A global topology map of the *Saccharomyces cerevisiae* membrane proteome. Proc Nat Acad Sci USA 103: 11142–11147

Klabunde T and Hessler G (2002) Drug design strategies for targeting G-protein-coupled receptors. Chembiochem: Eur J Chem Biol 3: 928–944

Krogh A, Larsson B, von Heijne G, Sonnhammer EL (2001) Predicting transmembrane protein topology with a hidden Markov model: application to complete genomes. J Mol Biol 305: 567–580

Letunic I, Copley RR, Schmidt S, Ciccarelli FD, Doerks T, Schultz J, Ponting CP, Bork P (2004) SMART 4.0: towards genomic data integration. Nucleic Acids Res 32: D142–D144

Lo Conte L, Ailey B, Hubbard TJ, Brenner SE, Murzin AG, Chothia C (2000) SCOP: a structural classification of proteins database. Nucleic Acids Res 28: 257–259

Lomize AL, Pogozheva ID, Lomize MA, Mosberg HI (2006a) Positioning of proteins in membranes: a computational approach. Protein Sci 15: 1318–1333

Lomize MA, Lomize AL, Pogozheva ID, Mosberg HI (2006b) OPM: orientations of proteins in membranes database. Bioinformatics 22: 623–625

Melén K, Krogh A, von Heijne G (2003) Reliability measures for membrane protein topology prediction algorithms. J Mol Biol 327: 735–744

Miller J (1972) Experiments in molecular genetics. Cold Spring Harbor, New York

Mitaku S, Ono M, Hirokawa T, Boon-Chieng S, Sonoyama M (1999) Proportion of membrane proteins in proteomes of 15 single-cell organisms analyzed by the SOSUI prediction system. Biophys Chem 82: 165–171

Möller S, Kriventseva EV, Apweiler R (2000) A collection of well characterised integral membrane proteins. Bioinformatics 16: 1159–1160

Murata K, Mitsuoka K, Hirai T, Walz T, Agre P, Heymann JB, Engel A, Fujiyoshi Y (2000) Structural determinants of water permeation through aquaporin-1. Nature 407: 599–605

Oomen CJ, van Ulsen P, van Gelder P, Feijen M, Tommassen J, Gros P (2004) Structure of the translocator domain of a bacterial autotransporter. EMBO J 23: 1257–1266

Orengo CA, Martin AM, Hutchinson G, Jones S, Jones DT, Michie AD, Swindells MB, Thornton JM (1998) Classifying a protein in the CATH database of domain structures. Acta Crystallogr D, Biol Crystallogr 54: 1155–1167

Ostermeier C and Michel H (1997) Crystallization of membrane proteins. Curr Opin Struct Biol 7: 697–701

Palczewski K, Kumasaka T, Hori T, Behnke CA, Motoshima H, Fox BA, Le Trong I, Teller DC, Okada T, Stenkamp RE, Yamamoto M, Miyano M (2000) Crystal structure of rhodopsin: a G protein-coupled receptor. Science 289: 739–745

Preusch PC, Norvell JC, Cassatt JC, Cassman M (1998) Progress away from 'no crystals, no grant'. Nat Struct Biol 5: 12–14

Punta M, Forrest LR, Bigelow H, Kernytsky A, Liu J, Rost B (2007) Membrane protein prediction methods. Methods 41: 460–474

Raman P, Cherezov V, Caffrey M (2006) The membrane protein data bank. Cell Mol Life Sci 63: 36–51

Rapp M, Drew D, Daley DO, Nilsson J, Carvalho T, Melén K, De Gier J, Von Heijne G (2004) Experimentally based topology models for *E. coli* inner membrane proteins. Protein Sci 13: 937–945

Sansom MSP, Scott KA, Bond PJ (2008) Coarse-grained simulation: a high-throughput computational approach to membrane proteins. Biochem Soc Trans 36: 27–32

Scott KA, Bond PJ, Ivetac A, Chetwynd AP, Khalid S, Sansom MSP (2008) Coarse-grained MD simulations of membrane protein-bilayer self-assembly. Structure 16: 621–630

Sigrist CJA, Cerutti L, Hulo N, Gattiker A, Falquet L, Pagni M, Bairoch A, Bucher P (2002) PROSITE: a documented database using patterns and profiles as motif descriptors. Brief Bioinform 3: 265–274

Tusnády GE and Simon I (2001) The HMMTOP transmembrane topology prediction server. Bioinformatics 17: 849–850

Tusnády G and Simon I (2009) Shedding light on transmembrane topology. In: Protein structure prediction: method and algorithms. Wiley Book Series on Bioinformatics (in press)

Tusnády G, Dosztányi Z, Simon I (2004) Transmembrane proteins in the protein data bank: identification and classification. Bioinformatics 20: 2964–2972

Tusnády GE, Dosztányi Z, Simon I (2005a) TMDET: web server for detecting transmembrane regions of proteins by using their 3D coordinates. Bioinformatics 21: 1276–1277

Tusnády G, Dosztányi Z, Simon I (2005b) PDB_TM: selection and membrane localization of transmembrane proteins in the protein data bank. Nucleic Acids Res 33: D275–D278

Tusnády GE, Kalmár L, Simon I (2008) TOPDB: topology data bank of transmembrane proteins. Nucleic Acids Res 36: D234–D239

White S and Wimley W (1999) Membrane protein folding and stability: physical principles. Ann Rev Biophys Biomol Struct 28: 319–365

Xu EW, Kearney P, Brown DG (2006) The use of functional domains to improve transmembrane protein topology prediction. J Bioinform Comput Biol 4: 109–123

Topology prediction of membrane proteins: how distantly related homologs come into play

Rita Casadio, Pier Luigi Martelli, Lisa Bartoli and Piero Fariselli

Biocomputing Group, Bologna Computational Biology Network, University of Bologna, Bologna, Italy

Abstract

The first atomic-resolution structure of a membrane protein was solved in 1985. After 25 years and 213 more unique structures in the database, we learned some remarkable biophysical features that thanks to computational methods help us to model the topology of membrane proteins (White 2009). However, not all the features can be predicted with statistically relevant scores when few examples are available (Oberai et al. Protein Sci 15: 1723–1734, 2006). Too often the notion that similar functions are supported by similar structures is expanded far behind the limits of a safe sequence identity value (>50%) to select templates for modeling the membrane protein at hand. To select proper templates we introduce a strategy based on the notion that remote homologs can have a role in determining the structure of any given membrane protein provided that the two proteins are co-existing in a cluster. Sequences are clustered in a set provided that any two sequences share a sequence identity value ≥40% with a coverage ≥90% after cross-genome comparison. This procedure not only allows safe selection of a putative template but also filters out spurious assignments of templates even when they are generally considered as the structure reference to a given functional family. The strategy also can play a role in indicating which membrane protein sets still would be worthwhile a structural investigation effort. Possibly when more membrane proteins will be available, the clustering system will allow fold coverage of the membrane protein universe.

1 Introduction

The prediction of membrane protein topology and that of different post-translational modifications (PTMs) are important problems of sequence annotation after genome sequencing. Membrane proteins play a fundamental role in cell biology and are among

Corresponding author: Rita Casadio, Biocomputing Group, Bologna Computational Biology Network, University of Bologna, 40126 Bologna, Italy (E-mail: casadio@biocomp.unibo.it)

the most addressed targets of pharmaceutical and life science research. They perform basic functions within the cell biological processes, including cell signaling, energy conservation and transformation, ion exchange and transport. Membrane proteins are difficult to study (Newby et al. 2009). They are inserted to a different extent into the lipid membrane phase of cells and subcellular compartments, exposing portions of various sizes to polar's outer and inner environments (Phillips et al. 2009).

Most of the presently available computational methods allow predicting two basic features of membrane proteins: (i) the location of transmembrane domains along the protein chain (topography) and the location of the N- and C-termini with respect to membrane plane (topology). Topologic models are sufficient in many instances to design experiments to determine the location of the inner and outer loops with respect to the membrane and concomitantly the number of transmembrane domains (Casadio et al. 2003).

A complementary source of information helpful to model membrane proteins is the knowledge of the presence in their sequence of special functional motifs such as signal peptides (SPs) and glycosylphosphatidylinisotol (GPI) anchors. These two relevant signals strongly influence the folding of the mature protein, both diminishing the length of the final sequence by promoting the cleavage of specific segments at the N-termini (SPs) and C-termini (GPI-anchors), respectively.

Membrane proteins fold as bundles of alpha helices (all-alpha membrane proteins) or as barrels of beta sheets to minimize their conformation energy within the lipid membrane phase. Most all-alpha helices comprise a large fraction of hydrophobic residues. Given the hydrophobic nature of SPs and GPI-anchors, erroneous helices can be assigned as SPs or GPI-anchors in a genome-wide analysis. For this reason, they should be properly detected and eventually removed before filtering the protein sequence with computational methods.

Another relevant constraint that can be exploited to test and evaluate the protein topology is the putative presence of disulfide bridges. Disulfide bridges are covalent bonds that link together two different cysteine residues close in the protein 3D space. This tie short-circuits the protein backbone and constrains the protein folding (Martelli et al. 2004).

Finally the protein function can help in many instances to select a template for modeling the sequence at hand even in the case of distantly related homologs and this is strictly correlated to the problem of functional annotation of membrane proteins.

2 From membrane protein sequence to topologic models

Historically membrane proteins have been always considered a specific class of proteins with peculiar chemico-physical properties. Indeed their partial or total insertion into the membrane phase generates two typical solvent exposure interfaces of the protein: one is polar and the other hydrophobic. This feature is considered a major cause of the difficulties reported in generating high-resolution crystals (Newby et al. 2009). However, structure modeling of membrane proteins can also be facilitated when the constraints that the membrane phase poses to the protein folding are taken into account. Several computational methods have been developed through the years with the purpose of recognizing from the sequence to which extent the protein can interact with the membrane. This allows highlighting which regions are eventually exposed to the polar solvent on either surface of the membrane. Methods starting from the protein sequence can predict the topography, the topology or both. In the following, we will briefly review the main underlying ideas that are popular within the scientific community and promote the development of different methods.

2.1 Datasets of membrane proteins

When implementing a computational method for predicting the topology of membrane proteins and assessing its performance, a very basic step is the selection of a dataset of proteins whose topology is well known. A dataset for the prediction of protein topology comprises possibly non-redundant protein chains with their annotations, in the form of residue-labels that identify transmembrane regions, inner loops, and outer loops, (and maybe reentrant loops). In the early days when it was possible to count on the fingers of one hand the number of high-resolution membrane proteins present in the Protein Data Bank (PDB; Fariselli et al. 1993), transmembrane annotation was indirectly obtained using low-resolution experiments to enrich the benchmark set (Jones et al. 1994). These indirect annotations (sometimes contradictory) highlighted loop regions in the cytoplasmic sides or in the extra-cytoplasmic sides and the transmembrane regions were routinely inferred using hydrophobic profiles. Low-resolution data were compiled to generate the first and widely used dataset of membrane protein annotations (Möller et al. 2001). When more membrane protein 3D structures were available, it turned out that the low-resolution annotations were not as good as previously thought (Chen et al. 2002).

Presently, although the number of unique membrane protein structure increased up to 214, the exact extent of the membrane bilayer surrounding the folded chains is not known. This poses uncertainty to the exact termini of the transmembrane regions. Several semiautomatic methods have been developed to face this problem (Tusnády et al. 2004; Lomize et al. 2006) leading also to different results. An alternative way of defining the transmembrane segments is to adopt the secondary structure annotations as derived from the 3D protein structure (Martelli et al. 2003).

63

2.2 Scoring the accuracy of different methods

Most popular scoring indexes are listed in the following and they help in determining to which extent methods are reliable. However, a direct comparison of different implementations is often hampered by the adoption of different training/testing sets. Nevertheless, a good suggestion is to check in the original paper if these indexes have been correctly evaluated on a training/testing set with very low redundancy.

The Matthews correlation coefficient (MCC) for a given class s (in our case, membrane and non-membrane residue) is defined as

$$MCC(s) = \frac{p(s)n(s) - u(s)o(s)}{d(s)}, \tag{1}$$

where $d(s)$ is the factor

$$d(s) = [(p(s) + u(s))(p(s) + o(s))(n(s) + u(s))(n(s) + o(s))]^{\frac{1}{2}} \tag{2}$$

$p(s)$ and $n(s)$ are respectively the true positive and true negative predictions for class s, while $o(s)$ and $u(s)$ are the numbers of false positives and false negatives.

The sensitivity (Sn) for each class s is defined as

$$Sn(s) = \frac{p(s)}{p(s) + u(s)}. \tag{3}$$

The specificity (Sp) is the probability of correct predictions and it is expressed as follows:

$$Sp(s) = \frac{p(s)}{p(s) + o(s)}. \tag{4}$$

When much interest is focused on predicting the topology of membrane proteins, more stringent scoring indices such as Q_{ok} and Q_{top} can be adopted. These compute the transmembrane segment location (Q_{ok}) and the topology (Q_{top}) accuracy of a set comprising N_p proteins, respectively and are defined as

$$Q_{ok} = \frac{100 P_{ok}}{N_p}, \tag{5}$$

$$Q_{top} = \frac{100 T_{ok}}{N_p},$$ (6)

where P_{ok} is the number of proteins whose all transmembrane segments are correctly assigned. For each protein this is a binary measure, since the index assigns 1 (correct) or 0 (wrong), depending on the fact that a prediction meets both of the following conditions.

(i) The number of predicted segments equals the observed ones.
(ii) The overlap between the predicted and expected segments equals at least k residues. The number k can be fixed or determined by the length of the predicted and observed segments. T_{ok} is the number of proteins whose both topology (the loop sides) and transmembrane segment location are correct. Since T_{ok} is a subset of P_{ok}, Q_{top} is always lower than or equal to Q_{ok}.

2.3 Propensity scales versus machine learning-based methods

A basic feature that allows the prediction of membrane protein topology is the different residue composition of the regions exposed to polar and apolar environments. In particular, in all-alpha membrane proteins, hydrophobic residues are more abundant in alpha-helical segments than in loops. Differences also exist between the compositions of inner and outer loops, being the former richer in positively charged residues (von Heijne and Gavel 1988). All the prediction approaches try to catch these differences by assigning different propensity values to each residue in the sequence depending on the different portions of the protein. On the basis of experimental results and/or statistical considerations, the most simple methods assign propensity values to each residue type independently of the sequence context leading to a propensity scale for the 20 residues. The most popular experimental scales are the Kyte–Doolittle hydropathy scale (Kyte and Doolittle 1982), the Wimley–White free energy scales for water/interface and water/octanol partition (White and Wimley 1999), and the White–von Heijne free-energy scale for membrane insertion (Hessa et al. 2005). All the scales are only suited to discriminate transmembrane regions from polar loops. More sophisticated methods include topogenic signals and require the adoption of statistical approaches: several scales, each one describing the propensity of a residue for a particular portion of the protein (e.g., inner loop, helix inside, helix middle, helix outside, and outer loop), were derived from the analysis of proteins with known topology (Jones et al. 1994).

The assumption of independence of the propensity values from the sequence context requires a limited number of parameters and strongly limits the predictive power of the scale-based methods. Machine learning-based methods were developed to overcome the above-mentioned problems and specifically for their ability of extracting complex association rules from a training set containing known examples (Bishop 2006). These rules are stored into the numerical values assigned to the internal parameters of the machine learning-based methods and can be used further for predicting the context-dependent propensity values for all the residues of new sequences. Among the different machine learning tools, Neural Networks (NNs) are particularly suited to analyze the information contained in a sequence segment surrounding the residue to be predicted (Fariselli et al. 1993) and the evolutionary information contained in the sequence profile (Rost et al. 1995). More recently, Support Vector Machines (SVMs) have been adopted with a similar approach (Nugent and Jones 2009). Both NNs and SVMs, as well as scale-based methods, compute a propensity value for each residue in a sequence, without taking into account topologic constraints, such as the minimal length of a transmembrane region. For this reason they need to be post-processed with an optimization method, as described in Section 2.4.

Hidden Markov models (HMMs) are a widely used machine-learning approaches that are able to cast topologic constraints simply by designing a graphical model composed of states that represent the position of a residue in a protein structure and are connected by arrows that in turn represent the allowed transitions. After the training phase, emission probabilities for each residue are assigned to each state together with transition probabilities among these states. Different decoding algorithms have been developed for globally predicting the topology of a new sequence (Sonnhammer et al. 1998). HMMs have been widely used for predicting the membrane protein topology (Tusnády and Simon 2001; Martelli et al. 2002, 2003; Käll et al. 2004, 2005; Viklund and Elofsson 2004).

2.4 Methods for optimizing topologic models

The best performing predictors of membrane protein topology generally include tools to optimize the length of transmembrane segments. This is a historical need since the local residue propensity is not enough to fully exploit the putative protein folding. The first tools based on the classical hydropathy scales already defined rules to accept transmembrane segments on the basis of the maximal number of residues included in the region (Kyte and Doolittle 1982). In practice, the positive regions of the plots of the protein sequences are scored in terms of the value of the maximum peak, the area of the positive region and its length. Then some empirical rules are adopted to accept the positive regions as a transmembrane segment (or divide it into different transmembrane segments). Kyte and Doolittle implemented one of the first bioinformatic programs in C language to perform the kinds of computations described above (Kyte and Doolittle 1982, Appendix). A similar filtering technique was also implemented in the early machine-learning approaches, where the local residue scores were obtained with a NN system (Rost et al. 1995). However, segment optimization procedures require several ad hoc rules that did not always guarantee global optimization of the predictive scores.

A more rigorous approach was obtained by developing a dynamic programming algorithm that globally optimizes the sum of residue scores and concomitantly fulfills the requirement that all the predicted transmembrane segments have a length compatible with the bilayer crossing (Jones et al. 1994). A similar method was also implemented to improve the accuracy of NN predictors (Rost et al. 1996). Later these ideas were cast into a more widely applicable algorithm suited to optimize the global score with segment constraints and capable to filter outputs of different types of predictors (Fariselli et al. 2003b).

The optimization algorithms based on dynamic programing are very similar to the HMM decoding procedures, such as the Viterbi algorithm (Bishop 2006). For this reason, differently from NN- and SVM-based predictors, results out of HMM-based methods do not need to be post-processed with an optimization algorithm. HMMs automatically incorporate in their automaton grammar global constraints. This can be exploited by adopting the Viterbi or Posterior-Viterbi decodings to automatically assign predictions that are compatible with the membrane constraints (Fariselli et al. 2005).

Finally, the best performing methods available evaluate the topology prediction by optimizing a global score and by maintaining the constraints derived from the

Table 1. Available servers for the prediction of membrane protein topology

Name	Method	Single/ multiple sequence	Signal peptide prediction	URL	References
ENSEMBLE	NN and HMM	Multiple	No	gpcr.biocomp.unibo.it/predictors	Martelli et al. (2003)
HMMTOP	HMM	Single	No	www.enzim.hu/hmmtop	Tusnády and Simon (2001)
MEMSAT-SVM	SVM	Multiple	Yes	bioinf4.cs.ucl.ac.uk:3000/psipred	Nugent and Jones (2009)
PHDhtm	NN	Multiple	No	www.rostlab.org	Rost et al. (1996)
PHOBIUS	HMM	Single	Yes	phobius.sbc.su.se	Käll et al. (2004)
POLYPHOBIUS	HMM	Multiple	Yes	phobius.sbc.su.se	Käll et al. (2005)
PRODIV	HMM	Multiple	No	topcons.cbr.su.se	Viklund and Elofsson (2004)
SPOCTOPUS	NN and HMM	Multiple	Yes	octopus.cbr.su.se	Viklund et al. (2008)
TMHMM	HMM	Single	No	www.cbs.dtu.dk/services/TMHMM	Krogh et al. (2001)

dataset of membrane protein structures. In this way, the predicted models benefit by both local and global information (Table 1).

2.5 Single sequence versus multiple sequence profile

One of the historical events in the advancement of the protein structure prediction was the introduction of evolutionary information in the form of sequence profiles when protein secondary structure evaluation was the task at hand (Rost and Sander 1994). The key idea was the exploitation of the protein profile to represent the input sequence. Starting from a multiple sequence alignment containing the sequence of interest, the profile is obtained by computing the frequency of the different residues in the different alignment positions. Thus the sequence profile P of a protein sequence s consists of a $N \times L$ matrix whose N rows represent the 20 amino acidic residues and the L columns specify the protein sequence positions. Each profile entry $P(a, p)$ contains the frequency of the amino acid type a in the sequence position p.

The advantages of utilizing profiles in place of the multiple sequence alignments are: (i) the information on the residue conservation of each sequence position p can be given in input to a predictor as aggregate information using a 20-residue vector and (ii) the computational complexity is independent of the number of aligned sequences since the information is condensed in the frequency entries of the profile.

The sequence profile provided as input to a NN-based method increased the performance also for predicting the topology of membrane proteins (Rost and Sander 1994).

A step forward in the prediction of the membrane protein topology was the implementation of new HMMs allowing the emission of vectors instead of symbols (Martelli et al. 2002). By this, it is possible to exploit the information contained in the sequence profiles and at the same time to define the grammatical structure that the bilayer imposes to the topology of membrane proteins (Martelli et al. 2003).

It is worth noticing that also methods based on propensity scales can improve their performances by taking advantage of the evolutionary information (Fariselli et al. 2003a,b). For instance, it is possible to compute the propensity values of each sequence position using the sequence profile P and a propensity scale h by averaging along a sliding window of W residues as

$$H(i) = \frac{1}{|W|} \sum_{j \in W} \sum_{a \in N} P(a, i+j)\, h(a), \tag{7}$$

where N is the set of the 20 residues, $P(a, p)$ is the frequency of the residue type a in the sequence position p (the sequence profile) and $h(a)$ is its propensity value (*http://gpcr.biocomp.unibo.it/predictors/*; Psi Kyte-Doolittle within TRAMPLE).

Presently all the state-of-the-art methods for the prediction of the topology of membrane proteins (Table 1) achieve their performances by exploiting evolutionary information. However, the introduction of sequence profiles makes the methods dependent on the selected alignment method, and this increases the computational time required to generate a prediction. For each protein sequence to be predicted a corresponding sequence profile must be created by searching and aligning a large set of similar proteins. Although today computing sequence profiles can count on huge computational power and fast heuristic algorithms such as PSI-BLAST (Altschul and Koonin 1998), the waiting time required to annotate entire genomes can be significantly long.

2.6 Prediction of signal peptides and GPI-anchors

An important issue when predicting membrane protein topology starting from genomic sequences is that N-terminal SP as well as C-terminal peptides cleaved upon GPI-anchoring are often incorrectly predicted as transmembrane alpha-helices. To address this problem, both SPs and GPI propeptides can be predicted with specific tools and sequences are currently be preprocessed by deleting the cleaved segments. Predictors for SPs include SignalP (Bendtsen et al. 2004) and SPEPlip (Fariselli et al. 2003a,b), both based on NNs.

A recent approach incorporates the prediction of SPs in tools suited to membrane protein topology determination by means of HMMs (Käll et al. 2004; Viklund et al. 2008) and SVMs (Nugent and Jones 2009) (Table 1). Specific tools are also available for predicting mitochondrial and chloroplastic N-terminal target peptides, such as TargetP (Emanuelsson et al. 2000) and MITOPROT (Claros and Vincens 1996).

Despite the paucity of known examples to be included in the training sets, prediction of the C-terminal peptides cleaved upon GPI-anchoring is efficiently performed by FragAnchor (Poisson et al. 2007) and PredGPI (Pierleoni et al. 2008).

2.7 More methods are better than one: CINTHIA

Prompted by the results described above we decided to take advantage of the different state-of-the-art methods to better annotate all-alpha membrane proteins focusing on the human genome. The recently developed Consensus of International Transmembrane Helical Intelligent Annotators (CINTHIA) can be regarded for a whole human proteome as an annotation process performed by a metapredictor. The CINTHIA pseudocode is described in Fig. 1. CINTHIA is a very simple metapredictor that exploits the predictions made by ENSEMBLE_2.0, MEMSAT3,

```
For each sequence in the Human Proteome:
        Run PSI-BLAST to get alignments for the query sequence
        Predict SP, and GPI-anchor
        if Positively predicted as GPI-anchor:
             cut the sequence at the N-terminus
        if Positively predicted as Signal Peptide:
             cut the sequence at the C-terminus
        Predict the sequence with SP+GPI cuts with all TM-predictors(*)
        Predict the sequence using CINTHIA
        Update the DB
```

Fig. 1. The pseudocode of the CINTHIA procedure. (*) In the case of Polyphobius that includes signal peptide prediction, only the GPI-anchor is removed.

and PRODIV_0.92 (see Table 1 for references) and uses the MaxSubSeq algorithm (Fariselli et al. 2003a,b) to find the best scoring model. The prediction of important features including the presence of SPs, the presence of disulfide bonds, and the predicted topologic models with different methods are included in the process (Fig. 1). In the case of SPs and GPI anchors, when the predicted features are present, the corresponding segments are excluded by the process to equalize strings delivered to different predictors. The metapredictor then assigns optimal scores to a topologic model for any given sequence. CINTHIA is therefore the final step of the annotation process and it needs almost all previous methods to be evaluated. On a dataset consisting of 131 high-resolution protein structures CINTHIA outperforms the accuracy of the single methods (Table 2), providing a more synthetic and reliable annotation system of the all-alpha membranome of the human genome.

When the human genome is predicted, the distribution of the CINTHIA annotations follows the pattern shown in Fig. 2, with a huge number of proteins annotat-

Table 2. Accuracy of the different methods on a set of high-resolution proteins

Method	Topography ($Q_{ok}{}^a$)	Topology ($Q_{top}{}^a$)
ENSEMBLE 2.0	0.86	0.74
MEMSAT3	0.89	0.84
POLYPHOBIUS	0.82	0.66
PRODIV 0.92	0.86	0.74
TMHMM 2.0	0.76	0.60
CINTHIA	0.93	0.86

[a]Prediction indexes are defined in Section 2.2.

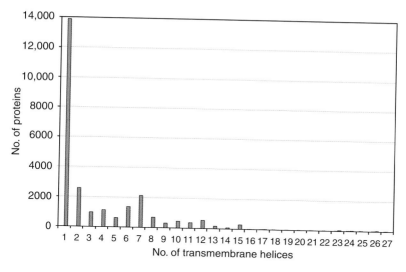

Fig. 2. Distribution of the number of proteins as function of the number of transmembrane helices predicted by CINTHIA in the human membranome.

ed with a single transmembrane spanning helix and a smaller peak that corresponds to proteins predicted with seven transmembrane regions.

2.8 A large-scale annotator of the human proteome: the PONGO system

The annotation efforts of the BIOSAPIENS European Network of Excellence (Juncker et al. 2009) have generated several distributed annotation systems (DAS) with the aim of integrating bioinformatics resources and annotating metazoan genomes (*http://www.biosapiens.info*). In this context, the PONGO DAS server (*http://pongo.biocomp.unibo.it*) provides the annotation on predictive basis for the all-alpha membrane proteins in the human genome, not only through DAS queries, but also directly using a simple web interface. We recently developed a new version of the PONGO system. This new version still maintains the functionalities of the previous browsable database version (Amico et al. 2006), and adds new features, including:

* CINTHIA, a consensus of the profile-based transmembrane predictors (see above).
* Predicted PTMs such as SPs, GPI-Anchors, and disulfide-bonds.

The new PONGO still adopts the previously developed technology defining its environment/web server/DAS server, which is based on a relational database containing all the data generated by the work package (Amico et al. 2006). PONGO can be queried through the web interface available at *http://pongo.biocomp.unibo.it* (or following a link from *www.biocomp.unibo.it*), or through the DAS client at EBI (*www. ebi.ac.uk*). The local DAS annotation server, administered by the Bologna Biocomputing Group at the Bologna University, is resident on the same machine.

The end-user can trace for each UniProt (*www.uniprot.org*) sequence of the human proteome whether the protein is or is not endowed with a SPs, a GPI-anchor, whether the sequence is or is not a membrane protein, and in this case its putative topology. The topologic model is computed by three predictors and also by CINTHIA on the basis of three profile-based predictors of transmembrane regions (see above). This allows to make a direct comparison among different predictors at the same time and to assess whether the expected results are or are not in agreement with the end-user experimental finding. Furthermore, the new PONGO displays the annotations of putative cysteines that can be involved into disulfide bonds (Fig. 3). By this, an

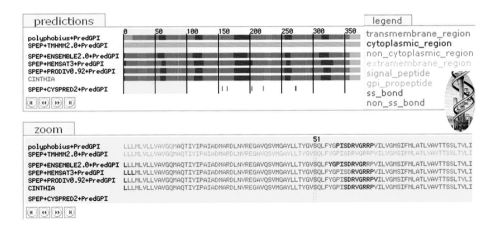

Fig. 3. A typical output of the PONGO annotation system (*pongo.biocomp.unibo.it*). See text for explanation on the annotation system.

Table 3.   PONGO annotates the human proteome

No. of human proteins in PONGO	92,186
% globular proteins	72
% proteins with a GPI anchor annotation	1
% proteins with a signal peptide annotation	18
% proteins with all-alpha transmembrane annotations	28

estimation of the total number of proteins in the human genome with a given feature can also be obtained. The data are shown in Table 3.

3 From membrane protein sequence to function and structure

In the previous section, we outlined the state-of-the-art methods to conventionally address the problem of membrane protein topology prediction. A still open question is as to which extent function can help in identifying the best topologic model. Most of the membrane proteins are usually clustered on the basis of general broad functions. According to this procedure, eventually common domains can be detected after modeling with HMMs and this can help in protein alignments of sequences with an overall very low pairwise sequence identity (*pfam.sanger.ac.uk*). When a "family" contains protein/s known with atomic resolution, these are regarded as a putative structural template for all the sequences within the set. As an example, all the proteins in the super family of G-protein-coupled receptors routinely inherited the structure of bacteriorhodopsin (Pebay-Peyroula et al. 1997) as long as the structure of bovine rhodopsin became available (Palczewski et al. 2000). This structure turned out to be somewhat different from that predicted on the basis of the bacteriorhodopsin template. Furthermore, since in the PDB membrane proteins of bacterial origins are generally more abundant than proteins from Eukaryotes, the eukaryotic target in the functional family is modeled considering the prokaryotic fold as the necessary template (*www.rcsb.org/pdb/home/home.do*). By now, several examples of proteins crystalized after modeling have indicated that this procedure may be risky (Casadio et al. 2002). For most of the membrane proteins clustered into conventional functional families (*pfam.sanger.ac.uk*), sequence identity is barely detectable and there is no a priori knowledge that can ensure that function conservation necessarily implies also structure conservation at a non-statistically significant level of sequence identity.

To cope with the problems discussed above and prompted by the necessity of large-scale genome annotation, we recently developed a non-hierarchical cluster-based procedure for gene functional and structural annotation. The method (the Bologna Annotation Resource, BAR) is based on a large-scale cross genome comparison of 599 genomes, including some 551 Prokaryotes and 48 Eukaryotes for a total of 2,624,555 protein sequences (Bartoli et al. 2009). BAR tested on some other 201 completely sequenced genomes was successfully able to improve the annotation process both in relation to structure and function (Bartoli et al. 2009). The clustering procedure that was adopted is based on the notion that any two protein sequences to belong to the same cluster share a sequence identity value ≥40% with a length of the aligned region normalized to the alignment length (coverage) ≥90%.

73

These stringent criteria ensure that in a cluster only sequences with a similar length are collected and that when a template/s is/are present the sequences are covered for most of their length. Furthermore, by assigning specific Gene Ontology (GO) terms (*www.geneontology.org*) to the clusters and by statistically validating them, the presence in a cluster of a given sequence ensures that the sequence, when possible, inherits also a function. This function is detailed by the corresponding GO terms. Clusters can contain 3D structures with structural domains classified according to SCOP (*scop.mrc-lmb.cam.ac.uk/scop/*).

With BAR, four main levels of annotations are possible with different alternatives for a total of 11 fine-tuned levels of annotations: (i) cluster-specific GO; i1: PDB/SCOP monodomain; i2: PDB/SCOP multidomain; i3: PDB without SCOP annotation; i4: without PDB; (ii) GO terms with *P*-value greater than the selected threshold (*P*-value >0.001); i5: PDB/SCOP monodomain; i6: PDB/SCOP multidomain; i7: PDB without SCOP annotation; i8: without PDB; (iii) without GO; i9: PDB/SCOP monodomain; i10: PDB/SCOP multidomain; i11: PDB without SCOP annotation; and (iv) no annotation.

Two main results can be obtained with BAR: (i) analyzing the most represented genes among genomes, finding genes that are expressed in all the kingdoms and inferring common traits of the living machinery of all prokaryotic and eukaryotic organisms and (ii) annotating protein sequences and assigning functions and structures to sequences that do not share a high level of similarity with the annotated ones. In the following, we highlight how this can help the recognition of membrane protein folds.

3.1 Membrane proteins: how many with known functions and folds?

Considering sequences from 599 genomes with BAR we may evaluate that the percentage of membrane proteins with GO terms is about 30. Out of this some 2% sequences are also endowed with PDB structure/s in the clusters.

3.1.1 All-alpha membrane proteins

In BAR 89 clusters, comprising 28,320 protein sequences, have at least one all-alpha membrane structure that can be considered according to our procedure as the template/s of the cluster. As shown in Table 4, 27,953 membrane proteins (98.7% of the total) inherit a template 3D structure not directly linked to them by other annotation procedures such as those of UniProt. When present in the same cluster, templates are highly similar [root-mean square deviation (RSMD) within 0.2 nm]. Our results support the conclusion that by following the BAR clustering procedure, known folds can be adopted as templates for a much larger fraction of

74

Table 4. BAR annotation of all-alpha membrane proteins in 599 complete genomes

		GO	GO (P-value ≤ 0.001)	GO (P-value > 0.001)	– GO
PDB	No. of clusters	81 (91%)	73 (82%)	8 (9%)	8 (9%)
	No. of sequences	28,216 (99.6%) [direct: 298 (1%), inherited: 27,918 (98.6%)]	28,054 (99.1%) [direct: 291 (0.98%), inherited: 27,763 (98.12%)]	162 (0.5%) [direct: 7 (0.02%), inherited: 155 (0.48%)]	104 (0.4%) [direct: 10 (0.1%), inherited: 94 (0.3%)]
SCOP Mono-domain	No. of clusters	53 (60%)	48 (54.4%)	5 (5.6%)	5 (5.6%)
	No. of sequences	22,925 (81%) [direct: 162 (0.6%), inherited: 22,763 (80.4%)]	22,781 (80%) [direct: 160 (0.56%), inherited: 22,621 (79.4%)]	144 (1%) [direct: 2 (0.04%), inherited: 142 (0.96%)]	53 (0.19%) [direct: 5 (0.02%), inherited: 48 (0.17%)]
SCOP Multi-domain	No. of clusters	15 (17%)	14 (16%)	1 (1%)	–
	No. of sequences	4542 (16%) [direct: 102 (0.4%), inherited: 4440 (15.6%)]	4540 (16%) [direct: 100 (0.4%), inherited: 4440 (15.6%)]	2 [direct: 2, inherited: –]	–

membrane protein sequences than before. Furthermore, also a functional annotation procedure is derived and statistically validated. The most represented molecular function GO terms in the clusters with a PDB template are transmembrane transporter activity (GO:0022857), substrate-specific transmembrane transporter activity (GO:0022891), ion binding (GO:0043167), and ion transmembrane transporter activity (GO:0015075). The most functional annotated proteomes in terms of membrane proteins are *Homo sapiens*, *Danio rerio*, *Gasterosteus aculeatus*, and *Macaca mulatta*.

3.1.2 All-beta membrane proteins

In BAR, 25 clusters have at least one all-beta membrane 3D structure that can be regarded as a template. By this, some 1475 protein sequences from 226 prokaryotic organisms can also inherit a template (Table 5). Interestingly enough, no eukaryotic porin falls into any of the well-annotated prokaryotic clusters containing beta-barrel structures. Here the most represented GO molecular functions are transmembrane transporter activity (GO:0022857), substrate-specific transmembrane transporter activity (GO:0022891), ion binding (GO:0043167), and cation transmembrane transporter activity (GO:0008324). The most functionally annotated proteomes in relation to all-beta membrane proteins are *Pseudomonas aeruginosa*, *Escherichia coli*, and *Salmonella typhimurium*.

Table 5. BAR annotation for beta-barrel membrane proteins in 599 complete genomes

		GO	GO (*P*-value ≤ 0.001)	GO (*P*-value > 0.001)	– GO
PDB	No. of clusters	22 (88%)	21 (84%)	1 (4%)	3 (12%)
	No. of sequences	1415 (96%) [direct: 66 (4.5%), inherited: 1349 (91.5%)]	1381 (94%) [direct: 66 (4.5%), inherited: 1315 (89.5%)]	34 (2%) [direct: –, inherited: 34 (2%)]	60 (4%) [direct: 14 (1%), inherited: 46 (3%)]
SCOP Mono-domain	No. of clusters	14 (56%)	13 (52%)	1 (4%)	2 (8%)
	No. of sequences	900 (61%) [direct: 50 (3%), inherited: 850 (58%)]	866 (59%) [direct: 50 (3%), inherited: 816 (56%)]	34 (2%) [direct: –, inherited: 34 (2%)]	51 (3%) [direct: 12 (1%), inherited: 39 (2%)]
SCOP Multi-domain	No. of clusters	1 (4%)	1 (4%)	–	–
	No. of sequences	57 (4%) [direct: 2 (0.3%), inherited: 55 (3.7%)]	57 (4%) [direct: 2 (0.3%), inherited: 55 (3.7%)]	–	–

3.2 What do BAR clusters contain?

In anyone of the clusters containing membrane protein structures, sequences with three threshold values of identity to the templates can be found: (i) very high values (≥50%). The high coverage of our clustering procedure ensures that when different templates are available from different sequences RMSD values of different templates coexisting in the same cluster are high; (ii) many sequences are endowed with identity values ≥40%. Also in this case building by comparison can be safely performed and the results can be obtained with any other method based on homology search; and (iii) the cluster contain sequences that are <40% identical to the sequence template/s (this may vary from 12% to 100% of the sequences in the cluster). In this case sequences with an identity much lower than 30% to the template are clustered together and can be regarded as distantly related homologs. By this, in spite of the very low sequence identity value, the remote homolog inherits both a structure and specific GO term/s. Furthermore our stringent clustering criteria also ensure high overlapping between the template and the target. Some clusters, containing membrane proteins endowed with structural motifs very difficult to predict with conventional predictors of topology, are detailed below (Bartoli et al. 2009).

3.2.1 The cluster of glyceroporins

Membrane proteins of the aquaporin family are highly selective for the permeation of specific small molecules with the exclusion of ions and charged solutes. Their structure is endowed with a unique feature in that two helices meet at their

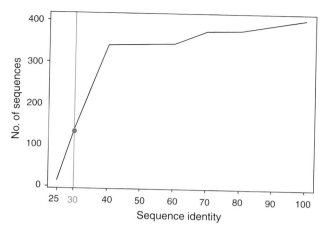

Fig. 4. Cumulative distribution of the number of sequences as a function of the sequence identity in the cluster of glyceroporins. The cluster contains 404 sequences from 256 organisms (226 Prokaryotes; 30 Eukaryotes), among which 9 (from *E. coli* and *Shigella sonnei*) are endowed with a PDB structure. Each sequence within the cluster was globally aligned to each of the 9 structural templates and the maximum value of sequence identity was considered. One hundred and thirty-seven sequences (34%) share <30% identity to a structural template and can therefore be modeled.

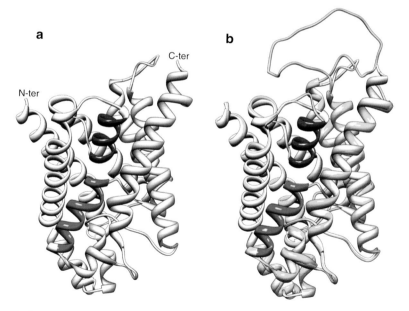

Fig. 5. 3D model of the glycerol uptake facilitator protein from *Streptococcus thermophilus*. (**a**) The glycerol uptake facilitator protein from *E. coli* (PDB ID: 1FX8) is the template. (**b**) Superimposition of the model of the Glycerol uptake facilitator protein from *Streptococcus thermophilus* (yellow) on the *E. coli* template (gray). Re-entrant alpha helices are shown in red (TMH3) and in blue (TMH7). N- and C-termini of the template face the cytoplasm. The RMSD value between the model and the template is 0.7 Å. Molecular models are visualized with UCSF Chimera (*www.cgl.ucsf.edu/chimera*).

77

N-terminal ends in the center of the membrane bilayer (re-entrant helices; Fu et al. 2000; Newby et al. 2008). Sequences cluster in a cluster that is named the cluster of glyceroporins since it contains nine sequences with PDB structures from *E. coli* and *Shigella sonnei*. Templates are structurally superimposable (RMSD = 0.13 nm). Interestingly enough in these clusters 226 sequences are from Prokaryotes and 30 from Eukaryotes. Clustering allows that some 137 sequences (34% of the total in the cluster) with <30% identity to the structural templates can also be modeled on the nine templates (Fig. 4). An example is given in Fig. 5, where the 3D model of the glycerol uptake facilitator protein from *Streptococcus thermophilus* was computed adopting as a template the counterpart from *E. coli* (sequence identity of the target to the template = 23%). The N- and C-termini of the template face the cytoplasmic side. Consequently the same topology is inherited by the target. The RMSD of the observed to the computed structure is 0.07 nm. Models are computed with Modeller (Eswar et al. 2008).

3.2.2 The cluster of multidrug transporter proteins (EmrE proteins)

EmrE is a multidrug transporter from *E. coli* that functions as a homodimer of a small four all-alpha transmembrane helices. Typical member of the small multidrug resistance family, its membrane topology was controversial till recent X-ray structural details that support a dual topology model of the dimeric functional unit

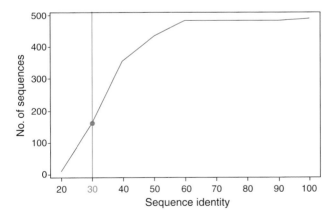

Fig. 6. Cumulative distribution of the number of sequences as a function of the sequence identity in the cluster of multidrug transporter proteins (EmrE proteins). The cluster contains 490 sequences from 280 prokaryotic organisms, among which 2 (from *E. coli*) are endowed with a PDB structure. Each sequence within the cluster was globally aligned to each of the two structural templates and the maximum value of sequence identity was considered. 161 sequences (29%) share <30% identity to a structural template and can therefore be modeled.

4.3.1 Tertiary structure evaluation metrics

The two measures used to evaluate tertiary predictions were root-mean square deviation (RMSD) and global distance test total score (GDT_TS). The latter has been used as the primary numeric measure in recent critical assessment of methods of protein structure prediction (CASP) experiments (Zemla et al. 2001; Moult et al. 2005). The TM notation is used as a subscript to indicate that the measure is calculated on only the TM segments of the true structure compared to the model.

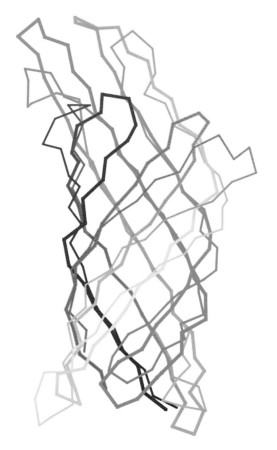

Fig. 3. C_α **trace of the lowest energy prediction for protein 1QJ8 aligned with the crystal structure.** Crystal structure trace is thicker and both traces are colored from cold (blue) to hot (red), from N- to C-termini. Calculated on all residues, the GDT_TS is 52.0 and the RMSD is 5.5 Å. Calculated on the curated TM segments only, the GDT_TS is 69.9 and the RMSD is 3.6 Å.

4.3.2 Prediction results

The best prediction, in terms of the GDT_TS and RMSD on the whole structure is made on the protein with the second highest β-contact recall: 1QJP. The β-contact recall is 0.65, the GDT_TS is 57.3 and RMSD is 4.3 Å. The GDT_TS_{TM} is 68.3 and $RMSD_{TM}$ is 3.0 Å. The next best whole structure predictions are for proteins 1QJ8 (52.0, 5.5 Å), 1PRN (50.0, 7.1 Å), and 1E54 (49.3, 7.7 Å). Figure 3 presents an image showing the predicted structure for 1QJ8 superimposed on the true structure. For several proteins the GDT_TS_{TM} results are strong. For proteins 1QJ8, 1QJP, 1PRN, 1I78, 1E54, 2OMF, and 1FEP the GDT_TS_{TM} is greater than 60.0. These predictions correspond to correct topology predictions and high β-contact recall when compared to the other predictions. The significantly lower GDT_TS and higher RMSD scores on the whole structures reflect the difficulty of modeling long loop regions and core domains folded inside the larger proteins.

The worst whole structure and TM segment predictions are made on proteins 1A0S and 2MPR, both of which have true strand counts of 18, but are modeled using 16-stranded templates because of incorrect secondary structure topology predictions. Additionally, the locations of multiple strands in the 2POR prediction are incorrect resulting in an incorrect topology according to the TOP measure. The worst whole structure and TM segment prediction for a protein with correct topology prediction was made on the 10-stranded protein 1K24. The topology is correct using the TOP measure; however, the locations of the sixth and seventh strands are off by seven residues. Using a slightly stricter standard for topology assessment, this prediction would be considered an incorrect topology. From these results it is clear that the correct topology is necessary to build a reasonable tertiary model.

The detailed tertiary structure prediction results for each protein in *SetPRED-TMBB* are available in table 4 of Randall et al. (2008).

4.3.3 Self-consistency results

To evaluate the self-consistency of TMBpro the curated secondary structure and true β-contacts were provided as input to the program. The performance was assessed both allowing and disallowing the inclusion of the native template among the available templates, and the results are displayed in the rightmost section of table 4 in Randall et al. (2008). When the native template is included, TMBpro always recovers the true structure. When the native template is not included, the $RMSD_{TM}$ results range from 1.5 to 4.5 Å. For 12 of 14 predictions, the $RMSD_{TM}$ is less than 2.8 Å. The only two exceptions are proteins 2FCP, with an $RMSD_{TM}$ of 3.5 Å, and 1QD5, with an $RMSD_{TM}$ of 4.5 Å. At 723 residues 2FCP is one of the longest proteins in the set, so a slightly higher error is not surprising. 1QD5 is only 269 residues, but contains an irregular bulge in the first strand that is not present in its only available template (1TLY).

5 Discussion

Recently, tertiary structure models predicted by TMBpro have been used to analyze mitochondrial porins in *Trypanosoma brucei* (Singha et al. 2009) and to compare channel characteristics between porin families of the marine cyanobacterium *Prochlorococcus* (current work with Dr. Adam Martiny of UC Irvine). The results of the latter project and the benchmark results of TMBpro trained on an updated set of non-redundant TMB proteins will be reported elsewhere. The updated version of TMBpro will be trained using the automatically calculated TM annotations from the OPM (Lomize et al. 2006) or PDB_TM (Tusnády et al. 2004, 2005b). Additional features to improve TMBpro's usability will include (1) an advanced interface that allows users to submit their own secondary structure definitions, fix the number of TM segments, and select specific structural templates for model building and (2) inclusion of confidence estimates for each residue in the model and the model as a whole in the output returned to the user.

TMB proteins have clear biological and medical relevance. Due to their importance and the difficulty of experimentally determining their structures, accurate tertiary structure prediction of TMB proteins is an important task for the protein structure prediction community. Traditional homology modeling methods will perform well if the target protein is similar enough to a solved protein to create a quality alignment; however, for the vast majority of putative TMB proteins traditional homology modeling will fail. The construction rules TMB proteins follow provide a greatly reduced search space compared to the globular protein structure prediction problem. In this chapter the TMBpro methodology for predicting secondary structure, β-contacts, and tertiary structure of TMB proteins was described. The TMBpro prediction server is freely available as part of the Institute for Genomics and Bioinformatics suite of prediction tools at: *http://www.igb.uci.edu/servers/psss.html*.

References

Altschul S, Madden T, Schaffer A, Zhang J, Zhang Z, Miller W, Lipman D (1997) Gapped BLAST and PSI-BLAST: a new generation of protein database search programs. Nucleic Acids Res 25: 3389–3402

Bagos P, Liakopoulos T, Spyropoulos I, Hamodrakas S (2004a) A hidden markov model method, capable of predicting and discriminating beta-barrel outer membrane proteins. BMC Bioinform 5: 29

Bagos P, Liakopoulos T, Spyropoulos I, Hamodrakas S (2004b) PRED-TMBB: a web server for predicting the topology of beta-barrel outer membrane proteins. Nucleic Acids Res 32: W400–W404

Bagos P, Liakopoulos T, Hamodrakas S (2005) Evaluation of methods for predicting the topology of beta-barrel outer membrane proteins and a consensus prediction method. BMC Bioinform 6: 7

Baldi P and Pollastri G (2003) The principled design of large-scale recursive neural network architectures-DAG-RNNs and the protein structure prediction problem. J Mach Learn Res 4: 575–602

Baldi P, Brunak S, Chauvin Y, Andersen C, Nielsen H (2000) Assessing the accuracy of prediction algorithms for classification: an overview. Bioinformatics 16: 412–424

Berman H, Westbrook J, Feng Z, Gilliland G, Bhat T, Weissig H, Shindyalov I, Bourne P (2000) The protein data bank. Nucleic Acids Res 28: 235–242

Bigelow H and Rost B (2006) PROFtmb: a web server for predicting bacterial transmembrane beta barrel proteins. Nucleic Acids Res 34: W186–W188

Bigelow H, Petrey D, Liu J, Przybylski D, Rost B (2004) Predicting transmembrane beta-barrels in proteomes. Nucleic Acids Res 32: 2566–2577

Casadio R, Fariselli P, Martelli P (2003) In silico prediction of the structure of membrane proteins: is it feasible? Brief Bioinform 4: 341–348

Cheng J and Baldi P (2005) Three-stage prediction of protein beta-sheets by neural networks, alignments, and graph algorithms. Bioinformatics 21 (Suppl 1): i75–i84

Cheng J, Sweredoski M, Baldi P (2005) Accurate prediction of protein disordered regions by mining protein structure data. Data Mining Knowl Discov 11: 213–222

Cheng J, Saigo H, Baldi P (2006a) Large-scale prediction of disulphide bridges using kernel methods, two-dimensional recursive neural networks, and weighted graph matching. Proteins 62: 617–629

Cheng J, Sweredoski M, Baldi P (2006b) DOMpro: protein domain prediction using profiles, secondary structure, relative solvent accessibility, and recursive neural networks. Data Mining Knowl Discov 13: 1–10

Diederichs K, Freigang J, Umhau S, Zeth K, Breed J (1998) Prediction by a neural network of outer membrane beta-strand protein topology. Protein Sci 7: 2413–2420

Fariselli P, Martelli P, Casadio R (2005) A new decoding algorithm for hidden Markov models improves the prediction of the topology of all-beta membrane proteins. BMC Bioinform 6: S12

Garrow A, Agnew A, Westhead D (2005) TMB-Hunt: an amino acid composition based method to screen proteomes for beta-barrel transmembrane proteins. BMC Bioinform 6: 56

Gromiha M and Suwa M (2005) A simple statistical method for discriminating outer membrane proteins with better accuracy. Bioinformatics 21: 961–968

Gromiha M, Majumdar R, Ponnuswamy P (1997) Identification of membrane spanning beta strands in bacterial porins. Protein Eng 10: 497–500

Gromiha M, Ahmad S, Suwa M (2004) Neural network-based prediction of transmembrane beta-strand segments in outer membrane proteins. J Comput Chem 25: 762–767

Gromiha M, Ahmad S, Suwa M (2005) TMBETA-NET: discrimination and prediction of membrane spanning beta-strands in outer membrane proteins. Nucleic Acids Res 33: W164–W167

Jackups R and Liang J (2005) Interstrand pairing patterns in beta-barrel membrane proteins: the positive-outside rule, aromatic rescue, and strand registration prediction. J Mol Biol 354: 979–993

Jacoboni I, Martelli P, Fariselli P, Pinto VD, Casadio R (2001) Prediction of the transmembrane regions of beta-barrel membrane proteins with a neural network based predictor. Protein Sci 10: 779–787

Kabsch W and Sander C (1983) Dictionary of protein secondary structure: pattern recognition of hydrogen-bonded and geometrical features. Biopolymers 22: 2577–2637

Koebnik R, Locher K, Gelder PV (2000) Structure and function of bacterial outer membrane proteins: barrels in a nutshell. Mol Microbiol 37: 239–253

Liu Q, Zhu Y, Wang B, Li Y (2003) A HMM-based method to predict the transmembrane regions of beta-barrel membrane proteins. Comput Biol Chem 27: 69–76

Lomize M, Lomize A, Pogozheva I, Mosberg H (2006) OPM: orientations of proteins in membrane database. Bioinformatics 22: 623–625

Martelli P, Fariselli P, Krogh A, Casadio R (2002) A sequence-profile-based hmm for predicting and discriminating beta barrel membrane proteins. Bioinformatics 18: S46–S53

Moult J, Krzysztof F, Rost B, Hubbard T, Tramontano A (2005) Critical assessment of methods of protein structure prediction (CASP) – Round 6. Proteins 61 (Suppl 7): 3–7

Natt N, Kaur H, Raghava G (2004) Prediction of transmembrane regions of beta-barrel proteins using ANN- and SVM-based methods. Proteins 56: 11–18

Oberai A, Ihm Y, Kim S, Bowie J (2006) A limited universe of membrane protein families and folds. Protein Sci 15: 1723–1734

Ou YY, Chen SA, Gromiha MM (2010) Prediction of membrane spanning segments and topology in beta-barrel membrane proteins at better accuracy. J Comput Chem 31: 217–223

Park K, Gromiha M, Horton P, Suwa M (2005) Discrimination of outer membrane proteins using support vector machines. Bioinformatics 21: 4223–4229

Paul C and Rosenbusch J (1985) Folding patterns of porin and bacteriorhodopsin. EMBO J 4: 1593–1597

Pollastri G and Baldi P (2002) Prediction of contact maps by GIOHMMs and recurrent neural networks using lateral propagation from all four cardinal corners. Bioinformatics 18: S62–S70

Pollastri G, Przybylski D, Rost B, Baldi P (2002) Improving the prediction of protein secondary structure in three and eight classes using recurrent neural networks and profiles. Proteins 47: 228–235

Randall A, Cheng J, Sweredoski M, Baldi P (2008) TMBpro: secondary structure, beta-contact and tertiary structure prediction of transmembrane beta-barrel proteins. Bioinformatics 24: 513–520

Remaut H, Tang C, Henderson NS, Pinkner JS, Wang T, Hultgren SJ, Thanassi DG, Waksman G, Li H (2008) Fiber formation across the bacterial outer membrane by the chaperone/usher pathway. Cell 133: 640–652

Schulz G (2000) Beta-barrel membrane proteins. Curr Opin Struct Biol 10: 443–447

Simons KT, Kooperberg C, Huang E, Baker D (1997) Assembly of protein tertiary structures from fragments with similar local sequences using simulated annealing and Bayesian scoring functions. J Mol Biol 268: 209–225

Singha UK, Sharma S, Chaudhuri M (2009) Downregulation of mitochondrial porin inhibits cell growth and alters respiratory phenotype in *Trypanosoma brucei*. Eukaryot Cell 8: 1418–1428

Skolnick J, Kolinski A, Ortiz A (1997) Monsster: a method for folding globular proteins with a small number of distance restraints. J Mol Biol 265: 217–241

Tamm L, Arora A, Kleinschmidt J (2001) Structure and assembly of beta-barrel membrane proteins. J Biol Chem 276: 32399–32402

Tamm L, Hong H, Liang B (2004) Folding and assembly of beta barrel membrane proteins. Biochim Biophy Acta 1666: 250–263

Tusnády GE, Dosztányi Z, Simon I (2004) Transmembrane proteins in the protein data bank: identification and classification. Bioinformatics 20: 2964–2972

Tusnády GE, Dosztányi Z, Simon I (2005a) PDB_TM: selection and membrane localization of transmembrane proteins in the protein data bank. Bioinformatics 33: D275–D278

Tusnády GE, Dosztányi Z, Simon I (2005b) TMDET: web server for detecting transmembrane domains by using 3D structure of proteins. Bioinformatics 21: 1276–1277

Waldispühl J, Berger B, Clote P, Steyaert J (2006a) Predicting transmembrane beta-barrels and inter-strand residue interactions from sequence. Proteins 65: 61–74

Waldispühl J, Berger B, Clote P, Steyaert J (2006b) transFold: a web server for predicting the structure and residue contacts of transmembrane beta-barrels. Nucleic Acids Res 34: W189–W193

Waldispühl J, O'Donnell CW, Devadas S, Clote P, Berger B (2008) Modeling ensembles of transmembrane beta-barrel proteins. Proteins 71: 1097–1112

Wallin E and von Heijne G (1998) Genome wide analysis of integral membrane proteins from eubacterial, archaean, and eukaryotic organisms. Protein Sci 7: 1029–1038

Welte W, Weiss M, Nestel U, Weckesser J, Schiltz E, Schulz G (1991) Prediction of the general structure of OmpF and PhoE from the sequence and structure of porin from *Rhodobacter capsulatus*. Orientation of porin in the membrane. Biochim Biophys Acta 1080: 271–274

Wimley W (2002) Toward genomic identification of beta-barrel membrane proteins: composition and architecture of known structures. Protein Sci 11: 301–312

Wimley W (2003) The versatile beta-barrel membrane protein. Curr Opin Struct Biol 13: 404–411

Yooseph S, Sutton G, Rusch DB, Halpern AL, Williamson SJ, Remington K, Eisen JA, Heidelberg KB, Manning G, Li W, Jaroszewski L, Cieplak P, Miller CS, Li H, Mashiyama ST, Joachimiak MP, van Belle C, Chandonia JM, Soergel DA, Zhai Y, Natarajan K, Lee S, Raphael BJ, Bafna V, Friedman R, Brenner SE, Godzik A, Eisenberg D, Dixon JE, Taylor SS, Strausberg RL, Frazier M, Venter JC (2007) The Sorcerer II Global Ocean Sampling expedition: expanding the universe of protein families. PLoS Biol 5: e16

Zhai Y and Saier M (2002) The beta-barrel finder (BBF) program, allowing identification of outer membrane beta-barrel proteins encoded within prokaryotic genomes. Protein Sci 11: 2196–2207

Zhang T, Kolinski A, Skolnick J (2003) TOUCHSTONE: II a new approach to ab initio protein structure prediction. Biophys J 85: 1145–1164

Zemla A, Venclovas C, Fidelis K, Rost B (1999) A modified definition of sov, a segment-based measure for protein secondary structure prediction assessment. Proteins 34: 220–223

Zemla A, Venclovas C, Moult J, Fidelis K (2001) Processing and evaluation of predictions in CASP4. Proteins 45 (Suppl 5): 13–21

Multiple alignment of transmembrane protein sequences

Walter Pirovano, Sanne Abeln, K. Anton Feenstra and Jaap Heringa

Centre for Integrative Bioinformatics, VU University Amsterdam, Amsterdam, The Netherlands

Abstract

Multiple sequence alignment remains one of the most powerful tools for assessing evolutionary sequence relationships and for identifying structurally and functionally important protein regions. Membrane-bound proteins represent a special class of proteins. The regions that insert into the cell membrane have a profoundly different hydrophobicity pattern as compared with soluble proteins. Multiple alignment techniques employing scoring schemes tailored for sequences of soluble proteins are therefore in principle not optimal to align membrane-bound proteins. In this chapter we describe some of the characteristics leading transmembrane proteins to display differences at the sequence level. We will also cover computational strategies and methods developed over the years for aligning this special class of proteins, discuss some current bottlenecks, and suggest some avenues for improvement.

Abbreviations: TM, transmembrane; MSA, multiple sequence alignment; SP, sum of pairs (score); TC, total column (score).

1 Introduction

Over the past years, integral membrane proteins have received a great deal of attention. They carry out essential functions in many cellular and physiologic processes, such as signal transduction, cell–cell recognition, and molecular transport. Membrane proteins are likely to constitute 20–30% of all ORFs contained in genomes (Jones 1998; Wallin and von Heijne 1998). Unfortunately, the number of determined transmembrane (TM) structures in the PDB is still very low: 1.7% (1139 of more than 64,000) are TM (Tusnády et al. 2005). Despite a solid growth of the

---

Corresponding author: Jaap Heringa, Centre for Integrative Bioinformatics, VU University Amsterdam, 1081 HV Amsterdam, The Netherlands (E-mail: heringa@few.vu.nl)

number of membrane protein structures (White 2004), and a steadily increasing fraction in the total PDB (up from ~1.4% in 2003), their determination remains a difficult task, such that they will continue to lag behind the number of experimentally solved soluble protein structures.

Given the biomedical importance of TM proteins and the large and growing gap between the number of solved TM protein structures and the number of TM protein sequences, sequence analysis techniques are crucial. The simultaneous alignment of three or more nucleotide or amino acid sequences is one of the most common tasks in bioinformatics. Multiple sequence alignment (MSA) is an essential pre-requisite to many further modes of analysis into protein families such as homology modeling, secondary structure prediction, phylogenetic reconstruction, or the delineation of conserved and variable sites within a family. Alignments may be further used to derive profiles (Gribskov et al. 1987) or hidden Markov models (Bucher et al. 1996; Eddy 1998; Karplus et al. 1998) that can be used to scour databases for distantly related members of the family.

The automatic generation of an accurate multiple alignment is potentially a daunting task. Ideally, one would make use of an in-depth knowledge of the evolutionary and structural relationships within the family but this information is often lacking or difficult to use. General empiric models of protein evolution (Dayhoff et al. 1978; Henikoff and Henikoff 1992) are widely used instead, but these can be difficult to apply when the sequences are less than 30% identical (Sander and Schneider 1991). Furthermore, mathematically sound methods for carrying out alignments, using these models, can be extremely demanding in computer resources for more than a handful of sequences (Carrillo and Lipman 1988; Stoye 1998). To be able to cope with practical dataset sizes, heuristics have been developed that are used for all but the smallest datasets.

The most commonly used heuristic methods are based on the progressive alignment strategy (Hogeweg and Hesper 1984; Feng and Doolittle 1987; Thompson et al. 1994; Heringa 1999). The idea is to establish an initial order for joining the sequences, and to follow this order in gradually building up the alignment. Many implementations use an approximation of a phylogenetic tree between the sequences as a guide tree that dictates the alignment order.

Although appropriate for many alignment problems, the progressive strategy suffers from its greediness. Errors made in the first alignments during the progressive protocol cannot be corrected later as the remaining sequences are added in. Attempts to minimize such alignment errors have generally been targeted at global sequence weighting (Altschul et al. 1989; Thompson et al. 1994), where the contribution of individual sequences is weighted during the alignment process. However, such global sequence weighting schemes carry the risk of propagating rather than reducing error when used in progressive multiple alignment strategies (Heringa 1999).

2 Factors influencing the alignment of transmembrane proteins

Transmembrane regions have a modified amino acid composition and different conservation patterns as compared to soluble proteins. Most current MSA techniques have been built, and optimized, to align homologous soluble proteins. Even though many such techniques are still applicable to TM regions, yielding a somewhat lower alignment accuracy than for soluble proteins (Forrest et al. 2006), there are some specific differences that should be taken into consideration when creating a MSA for TM proteins.

The lipid environment of TM regions influences the folding properties of the protein backbone and side chains, which can be observed not only in a different amino acid composition, but also in different evolutionary substitutions and different structural conservation. The hydrophobic nature of lipid tails of the membrane does not allow for H-bonding with the backbone or side chains of the peptide as water does. Therefore, backbone H-bonds of a TM region are typically satisfied through secondary structure: membrane spanning helices or beta barrels. The amino acid composition of TM regions is predominantly hydrophobic. These conditions are similar to the buried regions of soluble protein. TM regions nevertheless show some characteristic quite different from buried regions.

Donnelly et al. (1993) noted that TM helices, though mostly hydrophobic, had an alternating pattern of conserved and non-conserved amino acids; the conserved amino acids form the core of the protein structure, while the non-conserved hydrophobic amino acids point out toward the lipids. Furthermore, regions facing other protein parts within the lipid are typically enriched in phenylalanine and in tyrosine (Langosch and Heringa 1998; Bordner 2009). Jones et al. (1994) noted that polar regions are highly conserved within TM regions; this may be explained by charges or H-bonds needing to be satisfied by the protein itself, making polar residues in the membrane highly specific.

2.1 Transmembrane substitution rates

Conventional scoring matrices such as PAM (Dayhoff et al. 1978) or BLOSUM (Henikoff and Henikoff 1992), routinely used for sequence retrieval and alignment, are therefore not optimal to align TM regions. Several groups have made attempts to capture the evolutionary trends specific to TM regions in an amino acid substitution matrix, e.g., the JTT matrix (Jones et al. 1994), the PHAT matrix (Ng et al. 2000), the asymmetric SLIM matrices (Müller et al. 2001) and the bbTM matrix specialized for TM beta-barrels (Jimenez-Morales et al. 2008). Substitution scores, s_{ij}, are generally based on the frequency of amino acid substitutions, q_{ij}, in a set of aligned homologous sequences, according to:

Table 1. Difference between the PHAT and BLOSUM substitution matrices, expressed as PHAT75-73 - BLOSUM62 (see text for details)

	C	F	L	W	V	I	M	H	Y	A	G	P	N	T	S	R	Q	D	K	E
C	**-7.1**																			
F	4.0	-2.7																		
L	-1.1	1.6	-1.8																	
W	-2.2	-2.0	-0.7	-4.9																
V	-1.1	0.4	-0.4	-0.2	-1.8															
I	-2.7	0.0	-0.9	-0.2	-1.3	-0.2														
M	-1.1	0.0	-0.9	-4.2	-0.4	2.7	-0.7													
H	-4.9	-1.1	-0.2	-0.7	-1.8	-1.8	-2.2	1.1												
Y	2.4	0.2	-1.1	-2.4	-2.7	-2.7	-1.1	0.7	3.1											
A	1.6	2.4	0.4	-0.2	1.6	2.0	0.4	-0.7	-0.7	-0.2										
G	2.9	2.9	4.9	-3.8	2.9	4.9	4.4	-2.2	1.3	1.6	2.0									
P	**-6.5**	0.2	-1.8	-1.4	-2.2	-0.2	-3.8	**-5.4**	-1.8	-2.7	-0.7	**6.3**								
N	2.9	4.4	1.3	0.2	1.3	1.3	0.9	4.2	**7.1**	0.9	-1.6	-2.2	**5.1**							
T	0.4	0.9	0.4	**-6.9**	0.0	0.4	2.0	-2.2	-0.7	0.0	2.4	-4.2	-1.6	**-5.3**						
S	3.6	0.9	0.9	-1.8	0.9	0.9	-1.1	-1.1	0.9	1.1	1.6	-2.7	-0.4	-0.4	1.4					
R	**-6.5**	-4.9	**-5.4**	-4.9	-4.9	-3.4	**-7.4**	**-6.2**	**-5.4**	**-7.4**	-3.8	**-6.9**	-4.7	**-7.4**	**-7.4**	4.0				
Q	-1.8	2.9	-0.7	**5.6**	-0.7	1.3	-1.6	3.1	2.0	-2.7	0.9	-2.7	3.1	-2.7	-1.6	**-5.1**	4.0			
D	-4.9	-1.8	-1.8	-2.9	-1.8	-1.8	-1.8	0.4	-0.2	-3.8	-1.1	**-5.8**	1.1	**-5.8**	**-6.2**	**-6.9**	0.0	**6.7**		
K	**-9.6**	-4.9	**-6.9**	**-6.5**	**-8.5**	-4.9	**-7.4**	**-5.8**	-2.2	**-8.9**	-3.8	-4.2	-3.1	**-7.4**	**-7.8**	**-5.6**	-3.6	**-5.8**	-2.2	
E	-2.9	-1.8	-1.8	-4.9	-3.8	-3.8	-3.8	-1.6	0.9	**-5.8**	-0.7	**-5.8**	0.0	**-5.8**	-4.7	**-9.4**	-2.4	**5.4**	**-8.2**	**8.7**

Negative values are on a gray background, largest scores (>5 or <-5) are in bold.

$$s_{ij} = \frac{1}{\lambda} \ln \left(\frac{q_{ij}}{f_i f_j} \right),$$
(1)

where λ is a constant, and f_i are the background frequencies of amino acids. For the JTT matrix the q_{ij} and f_i were both calculated from aligned TM regions. The PHAT matrix calculates the observed substitutions in a similar fashion, but takes the expected number of substitutions $(f_i f_j)$ from general hydrophobic areas; in this way the background expectancy is based on TM as well as buried regions from soluble proteins. PHAT was shown to outperform the JTT matrix on database homology searches; this may be rationalized through background noise in alignments being similar to hydrophobic regions (Ng et al. 2000).

Table 1 shows the major differences in substitution rates between TM regions and soluble proteins, by subtracting the normalized PHAT matrix from the normalized BLOSUM62 matrix (Henikoff and Henikoff 1992). It can be observed that the polar residues have large positive self-substitution scores, i.e., the substitution scores in PHAT are higher than those in BLOSUM62, indicating that polar residues are more conserved in TM regions. Table 1 also shows that hydrophobic residues are less conserved, and that proline is particularly conserved in TM regions. The high proline conservation may be explained by the special role of proline residues forming kinks in TM regions (von Heijne 1991).

2.2 Transmembrane alignment gaps

Not only amino acid substitution rates, but also amino acid insertions and deletions show different patterns in TM proteins. Generally speaking the TM regions are much more conserved than their interconnecting loops (e.g., Forrest et al. 2006). These connecting loops may be very long, change considerably in size between homologs and they also show great structural flexibility or variability. Long loops are quite typical for TM proteins, and may be used for fly-casting (Dafforn and Smith 2004) or possibly to prevent aggregation of the highly hydrophobic TM regions (Abeln and Frenkel 2008); however, they pose a particular problem for MSA techniques. Typically gap open penalties should be higher for TM regions, as is shown for the PRALINE-TM example below (Pirovano et al. 2008a).

3 Overview of TM MSA methods

Not many techniques have been developed to improve the alignment of TM proteins. The method STAM (Shafrir and Guy 2004) represents an early attempt to improve alignment accuracy by combining different substitution matrices. A more recent study by Forrest et al. (2006) reported that the use of a bipartite scheme (consisting of BLOSUM62 and PHAT) does not significantly improve membrane protein sequence alignments. They suggest that the previously reported progress is more likely to depend on the separation of the TM blocks or on the settings of specific gap penalties.

We have recently investigated the effects of incorporating TM-specific information into the multiple alignment tool PRALINE, dubbed PRALINE-TM (Pirovano et al. 2008a). This information is integrated in a "soft" way, compared to the STAM approach where TM segments are first chopped and then aligned separately. In the PRALINE-TM approach the choice of the matrix depends on consistent TM predictions over a column and is determined dynamically during the alignment procedure. By applying the PHAT substitution matrix on consistently predicted TM regions, we show that it is possible to significantly improve the alignment quality.

3.1 TM-aware multiple sequence alignment by the Praline method

The strategy adopted by Praline for TM protein alignment, includes three basic techniques: (1) profile pre-processing, (2) a bipartite alignment scheme, and (3) tree-based iteration of the alignment.

3.1.1 Profile pre-processing

The profile pre-processing strategy in the PRALINE method (Heringa 1999) is a position-specific weighting scheme aimed at incorporating into each sequence, trusted information from other sequences. As such, it works contrary to the early weighting schemes mentioned above (Altschul et al. 1989; Thompson et al. 1994) which attempt to upweight sequences according to their divergence. In principle it is a good idea to perform global weighting aimed at increasing the contribution of more distant sequences as they carry more information at each alignment position. However, when sequence weighting is used in progressive multiple alignment, the increased chance of mistakes when aligning distant sequences can well lead to error propagation (Heringa 2002). Vogt et al. (1995) compared local and global alignments of pair-wise sequences with a data bank of structure-based alignments (Pascarella and Argos 1992) and included a set of over 30 substitution matrices with optimized gap penalties.

The best global alignments were achieved with the Gonnet residue exchange matrix (Gonnet et al. 1992), resulting in 15% incorrect residue matching when sequences with 30% residue identity were aligned. The error rate quickly increased to 45% incorrect matches at 20% residue identity of the aligned sequences, and to 73%

error at 15% sequence identity. Rost (1999) stressed the same point and reported even higher pair-wise alignment error rates in the twilight zone (below 30% identity). These statistics clearly demonstrate that increasing the global weight for distant sequences is likely to lead to misalignment and error propagation during progressive multiple alignment. That is why the Praline pre-processing strategy tends to upweight "trusted" sequences. For each sequence, a multiple alignment is created by stacking other sequences (master–slave alignment) that score beyond a user-specified threshold after pair-wise alignment with the sequence considered: a low threshold would result in a pre-processed alignment for each sequence comprising many or all other sequences (where the chance for alignment error is large), while higher thresholds would allow the information from fewer sequences into the alignment (with lesser alignment error). For each of the thus formed pre-processed alignments, a profile is constructed. The PRALINE method then performs progressive multiple alignment using the thus constructed pre-processed profiles. Each input sequence is now represented by its associated pre-processed profile, which incorporates knowledge about other "trusted" sequences (in particular similar sequences) and comprises position-specific gap penalties. This enables increased matching of distant sequences and appropriate placement of gaps outside ungapped core regions during progressive alignment, thereby avoiding errors early on in the progressive alignment.

3.1.2 Bipartite alignment scheme

The PRALINE bipartite strategy for TM proteins was implemented following the scheme devised for the alignment of soluble protein sequences for which 3-state secondary structure, i.e., α-helix, β-strand, and coil, is delineated (Heringa 1999). The PRALINE-TM tool first predicts the TM topology for each input sequence,

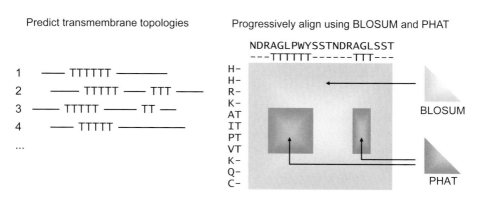

Fig. 1.   Overview of the PRALINE-TM bipartite alignment strategy.

using the state-of-the-art TM topology predictor Phobius (Käll et al. 2004). Next, the profile-scoring scheme simply applies TM-specific substitution scores from the PHAT matrix to consistently predicted TM positions, during the progressive alignment stage of PRALINE. In Fig. 1, an overview of this bipartite alignment strategy as implemented in PRALINE-TM is given. An alternative way of looking at the bipartite alignment strategy is to consider the TM info to be appended to residue types, thereby effectively doubling the number of residue types and quadrupling the size of the substitution matrix.

The current PRALINE profile-scoring scheme uses the following equation to score a pair of profile columns x and y:

$$S(x, y) = \sum_{i=1}^{20} \sum_{j=1}^{20} \alpha_i \beta_j M(i, j),$$
(2)

where α_i and β_j are the frequencies with which residues i and j appear in columns x and y, respectively, and $M(i, j)$ is the exchange weight for residues i and j provided by the selected substitution matrix M. By default profile columns are aligned using the BLOSUM62 matrix. Two profile columns will be matched using the PHAT matrix only in case each residue in the column is predicted to be member of a TM segment (see Fig. 1). This is done to guarantee that inconsistently predicted positions do not negatively influence the alignment quality. As a result, and contrary to the STAM method (Shafrir and Guy 2004), PRALINE-TM potentially allows TM segments to be aligned to non-TM segments (Pirovano et al. 2008a). The BLOSUM62 and PHAT substitution matrices are normalized using their diagonal elements as described by Abagyan and Batalov (1997). The "soft" bipartite scheme of Praline is less sensitive to errors in the delineation of the TM regions, as compared with "hard" bipartite schemes such as adopted in the STAM method. In fact, in the latter hard approaches the exact definition of the TM segment is critical because TM segments cannot be aligned with non-TM segments. Incorrectly delineated TM regions are likely to lead to misaligned TM and soluble segments, and no provision can be made for variable numbers of TM segments within families.

3.1.3 Tree-based consistency iteration

As a third and last step, the PRALINE-TM method employs an additional iterative strategy based on tree-dependent consistency iteration, which is similar to the tree-dependent strategy proposed by Hirosawa et al. (1995) and its implementation in the MUSCLE method (Edgar 2004). In this scenario, each edge of the phylogenetic (guide) tree is used to divide the alignment in two subalignments, which are successively realigned. The new alignment is retained only if an improved sum of pairs (SP) score is achieved. In the case of PRALINE-TM, this score is obtained by summing the substitution values of both the BLOSUM62 and PHAT matrix (depending on the TM topology of the amino acid pair). For the tree-based consistency strategy one iterative cycle implies that each edge of the tree is visited once. The maximum number of iterations is set to 20 by default to keep computations within bounds.

3.2 Bipartite MSA compared to standard MSA

It is important to know whether a bipartite scheme can improve alignment quality by including TM-specific information during the alignment procedure. Our recent work compares the PRALINE and (bipartite) PRALINE-TM strategies with three state-of-the-art TM topology predictors. The results are summarized in Table 2 and are based upon a standard progressive alignment strategy (which we refer to as "basic") to make a first comparison between an alignment method excluding and including the TM bipartite scheme. Accuracy is measured by comparing the alignment produced with the reference alignment. For BAliBASE, the BAliBASE "testing" program is provided that implements two scoring schemes: the SP score measures the fraction of correctly aligned residue pairs while the total column (TC) score expresses the fraction of correctly aligned columns. The TC score is the stricter of the two. (Note the distinction with the SP scoring used during alignment, which is based on a sum of substitution scores as explained in the previous section.)

A notable increase can be observed for all three TM predictors, albeit Phobius gives the best performance overall. Phobius has shown to be one of the most accurate TM topology predictors, especially on sequences that also contain a signal peptide (Käll et al. 2004; Jones 2007).

Independent contributions to the alignment quality coming from the PHAT matrix and TM-specific gap-open penalties were also investigated. The results in Fig. 2 clearly show that the combination of BLOSUM62/PHAT matrices yields optimal results, while using only BLOSUM62 or only PHAT does not, even when optimized

Table 2. Performance of the PRALINE and PRALINE-TM basic strategies (without pre-profiling) on reference set 7 of BAliBASE (at gap-open and gap-extension penalties of 15.0 and 1.0 for both the soluble and the transmembrane regions)

Method	SP score	TC score
PRALINE basic	0.646	0.231
PRALINE-TM basic – HMMTOP	0.679	0.264
PRALINE-TM basic – TMHMM	0.725	0.254
PRALINE-TM basic – Phobius	0.737	0.268

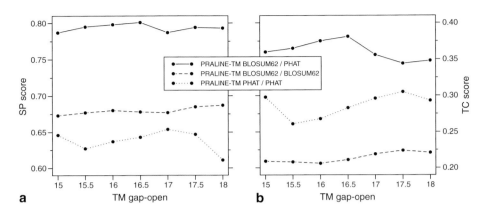

Fig. 2.   Contributions of the PHAT matrix and the gap-open penalty to the alignment quality.
The PRALINE-TM – Phobius method is evaluated at different TM gap-open penalties. In the first analysis (BLOSUM62/PHAT) the PHAT matrix is applied to the transmembrane regions. In the two control analyses either BLOSUM62 (BLOSUM62/BLOSUM62) or PHAT (PHAT/PHAT) is applied to the entire sequence. In (**a**) the average SP score is plotted, (**b**) the average TC score, as function of the TM gap-open penalty. Note that PRALINE-TM here includes the pre-processing strategy.

gap penalties are used (Pirovano et al. 2008a). Using only BLOSUM62, a small improvement can also be obtained by optimising the TM gap-open penalty; these influences however are much less pronounced than the use of the bipartite scheme.

The most striking observation to be made from both Table 2 and Fig. 2 is the positive effect on the alignment quality of the PHAT matrix applied to reliably predicted TM regions. In Table 2 the results are shown at an arbitrary gap-open penalty of 15.0 and gap-extension penalty of 1.0 for both the soluble and the TM regions. Figure 2 shows that the positive effect of using the PHAT matrix on TM regions is consistent over the whole range of gap-open penalties.

3.3 Comparing PRALINE-TM with non-TM MSA methods

Table 3 compares PRALINE-TM (gap-open penalty 15.0; TM gap-open penalty 16.5; pre-profile cut-off 11.0) with other widely used multiple alignment methods, which are designed for aligning soluble proteins. The default PRALINE method, i.e., "prepro" without TM information, with optimized parameter settings over this dataset is included for reference (gap-open penalty 15.0; pre-profile cut-off 8.5). Generally, methods reach average SP scores about 10% lower than for soluble proteins (Pirovano 2010). All methods reach SP scores that are twice as high as corresponding TC scores. PRALINE-TM achieves the highest SP score for two datasets and the highest TC score for four datasets. Concerning the averages overall eight

Table 3. Comparison between the PRALINE-TM method and five widely-used multiple alignment methods

Set	ClustalW	MUSCLE	MAFFT	ProbCons	PRALINE	PRALINE-TM
SP score						
7tm	0.847	0.836	0.835	**0.882**	0.816	0.860
Acr	0.906	**0.946**	0.937	0.935	0.930	0.936
Dtd	0.786	0.855	0.844	**0.877**	0.824	0.863
Ion	0.354	0.520	0.509	0.527	0.346	**0.543**
Msl	0.864	0.870	0.845	0.849	0.813	**0.874**
Nat	0.630	0.738	**0.766**	0.745	0.720	0.713
photo	0.887	0.902	**0.934**	0.913	0.915	0.933
ptga	0.461	0.551	**0.729**	0.716	0.404	0.683
Avg.	0.717	0.777	0.800	**0.806**	0.721	0.801
TC score						
7tm	0.410	0.340	0.320	0.410	0.310	**0.430**
Acr	0.580	0.670	0.620	0.670	**0.690**	0.620
Dtd	0.250	0.310	0.210	0.340	0.360	**0.390**
Ion	0.000	0.000	0.030	**0.090**	0.000	0.000
Msl	0.610	0.630	0.610	0.600	0.580	**0.660**
Nat	0.020	0.130	0.120	0.180	**0.220**	0.140
photo	0.490	0.460	0.550	0.490	0.570	**0.730**
ptga	0.010	0.060	**0.180**	0.150	0.000	0.080
Avg.	0.296	0.325	0.330	0.366	0.341	**0.381**

Individual and average SP and TC scores are given; for each set the best scoring method is highlighted in bold.

datasets, ProbCons slightly outperforms MAFFT and PRALINE-TM on the SP score. On the more critical TC score PRALINE-TM clearly scores best. ClustalW and MUSCLE score considerably lower on almost all datasets. The standard PRALINE method achieves a SP score comparable to ClustalW, but can be placed between MAFFT and ProbCons with respect to the TC score. Importantly, inclusion of TM information in PRALINE-TM improves overall SP and TC scores compared to standard PRALINE.

It should be mentioned that the PRALINE and PRALINE-TM methods were optimized on the TM dataset, whereas the other methods were run at default settings. It is noteworthy in this respect that an increased pre-profile cut-off value of 11.0 for PRALINE-TM worked best, in contrast to the cut-off value of 8.5 that was found to be optimal for soluble protein sequences. The "tighter" pre-profiles with fewer but more similar sequences yielded improved TM predictions, leading in turn to improved alignments. Concerning this training scenario, both MAFFT and ProbCons are relatively robust on TM sequences. Nonetheless, the results show clearly that a TM-based strategy can significantly improve the quality of TM protein sequence alignments, and should be considered a promising avenue for other applications as well.

4 Benchmarking transmembrane alignments

Owing to the underrepresentation of solved crystal TM structures in the PDB database, it is a difficult task to assess the quality of TM protein sequence alignments based on structural comparisons. To date, out of the many alignment benchmarks available for standard proteins, only BAliBASE (Bahr et al. 2001) has devoted a special reference set to TM proteins, ref7. This set contains eight accurately aligned TM families and consists of in total 435 sequences with an average length of 567 residues. The number of TM α-helices per sequence varies from 2 to 14.

For BAliBASE the SP and TC scores explained above are used. Other general benchmark sets use different names (the PREFAB "Q score" and SABmark f_D score are equivalent to the SP score), or different scores (SABmark f_M score is the total number of correctly aligned pairs from the test alignment, and is also stricter than the SP score). It is however not trivial to select the biologically most meaningful scoring scheme, as even gold-standard structure-derived alignments can show high variability (Pirovano et al. 2008b). In particular, this can be a serious problem for the strongly variable TM loop regions.

Despite the efforts that are being made to cover a significant part of the TM structure space (i.e., by taking families that display different evolutionary distances and a different number of TM helices), the number of sequences is rather low for robust testing of alignment methods. Particularly, only 15% of TM proteins in the PDB are porin-like β-barrel TM protein structures (Tusnády et al. 2005), so that a general benchmark set for β-TM alignments is currently lacking. Moreover, for a number of benchmark sequences the TM segments are determined from structure prediction, leading to a circular problem between TM alignment construction and prediction. The scarcity of reference data creates a problem for optimising any MSA technique for TM proteins, as overfitting of variables is unavoidable on such small and biased datasets.

The HOMEP dataset of homologous membrane proteins (Forrest et al. 2006) provides an alternative possibility for evaluating TM alignment accuracy. The set consists of 36 TM proteins with solved crystal structures which can be grouped into 11 SCOP families of similar topology. By making pair-wise structural superpositions within the families a set of 94 homologous comparative models has been compiled. For each pair of structures the TM location is determined both manually and using TMDET (Tusnády et al. 2005). This set has been the basis for the comparison of different sequence alignment algorithms as applied to membrane proteins by Forrest et al. (2006).

Other ways of testing the quality of TM alignments may be more indirect. For example, the performance of TM prediction methods can be used as a measure to evaluate alignment quality. It should be stressed that many of these methods use multiple alignments as an input for deriving prediction rules and paradoxically the quality of the alignment turns out to be a crucial factor here (Cserzö et al. 1994, 1997; Käll et al. 2005; Jones 2007). A scenario thus could be to select TM sequences for which corresponding crystal structures are available and retrieve for each of the homologous sequences using PSI-BLAST (Altschul et al. 1997). Subsequently, these sequence sets can be aligned by the competing alignment methods and serve as input for a TM prediction method. The prediction quality can be assessed using information on TM segment topologies and locations derived from the crystal structures, as routinely done for TM prediction method comparison (e.g., Nugent and Jones 2009).

4.1 Defining TM regions

A related and important issue is the definition of the membrane-bound region of the protein itself. In addition to the membrane-bound region, that strictly consists of the membrane-exposed "TM region" residues, the notion of membrane-spanning region (MSR) is often used (e.g., Möller et al. 2001; Forrest et al. 2006). The MSR is the secondary structure element, mostly α-helix but also β-strand, that crosses the membrane and potentially extends far outside the membrane region proper. The TM region therefore is a subset of the MSR. A definition based on X-ray crystallography data is also possible, provided sufficient electron density is observed that can be associated with the lipid region (Fyfe et al. 2001). Other experimental methods include proteolytic cleavage and chemical probe methods (e.g., Jennings 1989, for a review). In general, however, the boundaries of the lipid-associated region cannot be determined unambiguously (Möller et al. 2001).

The common method for defining the membrane-bound region of the protein is by the evaluation of the periodicity of hydrophobicity and conservation along the sequence, for example, the TMDET method and corresponding PDB_TM database (Tusnády et al. 2005). Buried residues, i.e., in the protein core, are on average as hydrophobic as membrane-exposed residues, but considerably more conserved (e.g., Donnelly et al. 1993). In addition, the membrane-associated regions are anchored transversally in the membrane by strongly conserved polar and/or positive residues that flank the hydrophobic region on either side. Propensity tables for core and head-group region location of amino acids, and also neural nets or Hidden Markov models (HMMs), are commonly used to identify or predict TM regions (Sonnhammer et al. 1998). Interestingly, for optimal training of a HMM-based model the gold-standard TM region definition used has to be adjusted; all segments were shortened on both ends by three residues, and the new "optimal" segment boundary was determined by the model itself (Sonnhammer et al. 1998; Krogh et al. 2001). Likewise, for evaluation of the performance of TM prediction methods generally an overlap between five and ten residues from the predicted TM segments with the 21 residues average length of the MSR is considered a successful prediction (Möller et al. 2001; Käll et al. 2004; Jones 2007). Please refer to Chapter 4 *Topology prediction of membrane proteins* by Rita Casadio et al. for a more extensive overview of this topic.

What therefore commonly remains undefined in the description of the TM prediction methods is whether it is the core-headgroup boundary, or the headgroup-solvent boundary that is actually predicted. For accurate TM MSA it is these boundaries that are of crucial importance to the successful selection of appropriate substitution matrices for the different regions. Moreover, the determination of the substitution scores depends critically on a well-defined demarcation of the regions of different composition and conservation patterns, especially since relatively little sequence data are available on which these scores can be based.

5 Applications for TM multiple alignments

Multiple sequence alignment is used for various problems wherein accurate alignments between two or more proteins are required. For TM proteins, there are in particular many applications that require accurate profiles of multiple homologous sequences. For example, the prediction of TM regions, TM topology prediction (e.g., Jones 2007; see also Chapter 4 *Topology prediction of membrane proteins* by Rita Casadio et al.), and the predictions of binding sites of TM proteins (Bordner 2009) all require accurate MSA; advances in MSA techniques for TM proteins may therefore also benefit these applications.

Owing to the lack of experimentally solved TM structures homology modeling is a particularly important application. MSA techniques may help to improve the alignment quality necessary for accurate homology modeling by adding information from other homologous sequences. Forrest et al. (2006) give a particularly helpful review on the homology modeling of TM proteins. They find that in general techniques developed for soluble proteins may also be used for TM proteins, although a slight drop in accuracy is observed as compared to soluble proteins. Surprisingly, the TM regions of the proteins are more accurately modeled than the interconnecting loops; this may be explained by the larger (structural) conservation of the TM regions, as they form the core of the TM domains. The transferability of methods developed for soluble proteins to TM proteins, such as secondary structure prediction, may be explained by the similar hydrophobic environment of buried regions in soluble proteins, and lipid exposed region in TM domains.

5.1 Homology searches of TM proteins

Alignments between (multiple) proteins are used for two distinct purposes to: (1) decide if two proteins are evolutionary related, a homology search, or (2) determine the most accurate alignment between (a set of) evolutionary-related proteins. The latter problem is usually addressed with MSA techniques. The two problems are strongly related, therefore techniques improving the alignment quality may also improve database search specificity and vice versa. For example, it has been shown that the PHAT substitution matrix, originally developed to improve homology searches, can also substantially improve the quality of multiple sequence alignments (Pirovano and Heringa 2008). This is not surprising as both problems are heavily dependent on substitution rates between amino acids. On a similar note, profiles created by homology searches may be used to improve MSA (Simossis et al. 2005).

Some particular issues have been observed for homology searches of membrane proteins; generally methods developed for soluble protein are used, such as PSI-BLAST (Altschul et al. 1997) or HMMER (Eddy 1998). However, it has been noted that false positive rates may go up considerably as compared to database searches for soluble proteins. Homology searches for TM proteins in general protein databases using BLAST may be improved by employing TM-specific substitution matrices (e.g., Ng et al. 2000; Müller et al. 2001; Jimenez-Morales et al. 2008). The compositional bias of TM proteins may be overcome with corrections to scoring (Schäffer et al. 2001), or substitution matrices (Altschul et al. 2005). However, to our knowledge, it has not been verified whether such compositional bias corrections indeed improve homology searches for TM proteins. In addition, homology searches may be improved by adding information about the predicted TM regions and their topology; this information is in particular valuable for the search of more remote homologs (Bernsel et al. 2008).

6 Current bottlenecks

A few significant problems concerning MSA method development are holding back progress in accurate alignment of TM proteins:

1. Benchmark sets are limited in quantity and quality, and are severely biased toward TM helices. This limits tuning of method parameters, accurate determination of substitution rates and proper cross-validated testing of the MSA methods.

117

2. Homology searches are known to yield relatively large numbers of false positives for TM proteins. It is unclear to what extent recent method developments have improved this issue. A thorough investigation of these false positive rates is necessary for current homology detection methods. Accurate homology detection, especially a low false positive rate, is crucial for the generation of high quality profiles to guide alignment and TM topology prediction.

3. The very definition of membrane regions remains highly ambiguous, both theoretically and practically. This poses potentially severe problems for accurate determination of substitution rates for the different regions (core, headgroup, and soluble). The optimal performance of a TM alignment method depends on a strict correspondence of the regions on which the substitution matrices have been calculated, and the predicted regions used in the bipartite alignment scheme.

7 Avenues for improvement

In contrast to the bottlenecks identified above, several feasible and practical adjustments are worth considering to improve alignment accuracy of TM proteins. We list them in approximate order of expected return on investment:

1. Current TM-specific substitution matrices have been determined about a decade ago. The increase in experimental TM protein data will allow better estimates for the exchange parameters, in particular for the β-barrel TM regions. Moreover the current substitution matrices were developed specifically for accurate homology searches, which will not necessarily yield the best results for alignment methods.

2. TM region prediction depends on accurate profiles, and is used in creating the TM multiple alignment. During the MSA procedure the profile quality progressively increases, which gives the opportunity to iteratively improve TM predictions and thereby the TM multiple alignment.

3. A known feature of TM loops is their flexibility and variation in length. The bipartite scheme implies separate gap penalty settings for the TM and non-TM regions, but does not distinguish between TM loops and non-TM domains.

4. Particular features of the TM regions, such as the "positive inside" rule, TM topology, location of the head groups which are also predicted by the TM prediction methods, are not explicitly used in current alignment schemes. For homology detection, inclusion of TM topology has significant positive effect (Bernsel et al. 2008). For example, penalties for misaligned regions may be included.

5. A more fine-grained approach for region-specific substitution matrices, as well as misalignment penalties mentioned above, could be beneficial. Particularly, the amino acid composition and substitution rates of β-barrels are known to differ from

current values for α-helical TM regions. Also (protein) buried versus lipid exposed TM residues show significantly different conservation patterns. Finally, the polar headgroup region imposes particular constraints on the amino acid composition.

6. Apparently, allowing predicted TM regions to be aligned with non-TM regions is necessary for accurate alignment of TM proteins; either due to inaccuracies in the TM prediction, flexibility in the length or due to number of TM regions between homologs. To optimize the TM to non-TM alignment one needs to correct for the compositional bias between these regions, for example, by non-symmetrical substitution matrices.

7. Further method improvements might come from more balanced and comprehensive benchmarks, not only containing alignments, but for instance also including validation of TM prediction methods based on different alignment inputs (from different alignment strategies).

8 Conclusions

MSA methods, trained on soluble proteins, in general do a reasonable job at aligning TM protein sequences; we approximate a difference of 10% in accuracy. Alignment quality can be improved significantly using a TM-specific substitution matrix and proper gap penalty settings, as shown by the PRALINE-TM example. In our view the improvement is mainly attributed by the fact that the bipartite scheme, using BLOSUM62 and PHAT, is applied in a flexible manner to undivided sequences during each step of the alignment procedure. Attempts where TM and soluble regions were aligned independently did not succeed in making significantly better alignments (Forrest et al. 2006). Strict gap penalty-settings for TM regions improve the overall performance; however, these effects should not be overestimated; the optimal TM gap-open penalty was only slightly higher than the standard penalty.

Overall we conclude that TM-awareness is an important concept for optimising MSA quality, yielding an increased performance of about 10%. However, none of the methods included here was able to align more than 40% of the reference alignment columns on average (TC score), so that further optimization remains a challenging task. Nevertheless, the difficulty of experimentally determining TM protein structures makes this a worthwhile effort.

References

Abagyan RA and Batalov S (1997) Do aligned sequences share the same fold? J Mol Biol 273: 355–368

Abeln S and Frenkel D (2008) Disordered flanks prevent peptide aggregation. PLoS Comput Biol 4: e1000241

119

Altschul SF, Carroll RJ, Lipman DJ (1989) Weights for data related by a tree. J Mol Biol 207: 647–653

Altschul SF, Madden TL, Schäffer AA, Zhang J, Zhang Z, Miller W, Lipman DJ (1997) Gapped BLAST and PSI-BLAST: a new generation of protein database search programs. Nucleic Acids Res 25: 3389–3402

Altschul SF, Wootton JC, Gertz EM, Agarwala R, Morgulis A, Schäffer AA, Yu YK (2005) Protein database searches using compositionally adjusted substitution matrices. FEBS J 272: 5101–5109

Bahr A, Thompson JD, Thierry JC, Poch O (2001) BAliBASE (Benchmark Alignment dataBASE): enhancements for repeats, transmembrane sequences and circular permutations. Nucleic Acids Res 29: 323–326

Bernsel A, Viklund H, Elofsson A (2008) Remote homology detection of integral membrane proteins using conserved sequence features. Proteins 71: 1387–1399

Bordner AJ (2009) Predicting protein–protein binding sites in membrane proteins. BMC Bioinform 10: 312

Bucher P, Karplus K, Moeri N, Hofmann K (1996) A flexible motif search technique based on generalized profiles. Comput Chem 20: 3–23

Carrillo H and Lipman D (1988) The multiple sequence alignment problem in biology. SIAM J Appl Math 48: 1073–1082

Cserzö M, Bernassau JM, Simon I, Maigret B (1994) New alignment strategy for transmembrane proteins. J Mol Biol 243: 388–396

Cserzö M, Wallin E, Simon I, Von Heijne G, Elofsson A (1997) Prediction of transmembrane alpha-helices in prokaryotic membrane proteins: the dense alignment surface method. Protein Eng 10: 673–676

Dafforn TR and Smith CJ (2004) Natively unfolded domains in endocytosis: hooks, lines and linkers. EMBO Rep 5: 1046–1052

Dayhoff MO, Schwart RM, Orcutt BC (1978) A model of evolutionary change in proteins. In: Dayhoff M (ed) Atlas of protein sequence and structure. National Biomedical Research Foundation, Washington, DC, pp 345–352

Donnelly D, Overington JP, Ruffle SV, Nugent JH, Blundell TL (1993) Modeling alpha-helical transmembrane domains: the calculation and use of substitution tables for lipid-facing residues. Protein Sci 2: 55–70

Eddy SR (1998) Profile hidden Markov models. Bioinformatics 14: 755–763

Edgar RC (2004) MUSCLE: a multiple sequence alignment method with reduced time and space complexity. BMC Bioinform 5: 113

Feng DF and Doolittle RF (1987) Progressive sequence alignment as a prerequisite to correct phylogenetic trees. J Mol Evol 25: 351–360

Forrest LR, Tang CL, Honig B (2006) On the accuracy of homology modeling and sequence alignment methods applied to membrane proteins. Biophys J 91: 508–517

Fyfe PK, McAuley KE, Roszak AW, Isaacs NW, Cogdell RJ, Jones MR (2001) Probing the interface between membrane proteins and membrane lipids by X-ray crystallography. Trends Biochem Sci 26: 106–112

Gribskov M, McLachlan AD, Eisenberg D (1987) Profile analysis: detection of distantly related proteins. Proc Natl Acad Sci USA 84: 4355–4358

Gonnet GH, Cohen MA, Benner SA (1992) Exhaustive matching of the entire protein sequence database. Science 256: 1443–1445

von Heijne G (1991) Proline kinks in transmembrane alpha-helices. J Mol Biol 218: 499–503

Henikoff S and Henikoff JG (1992) Amino acid substitution matrices from protein blocks. Proc Natl Acad Sci USA 89: 10915–10919

Heringa J (1999) Two strategies for sequence comparison: profile-preprocessed and secondary structure-induced multiple alignment. Comput Chem 23: 341–364

Heringa J (2002) Local weighting schemes for protein multiple sequence alignment. Comput Chem 26: 459–477

Hirosawa M, Totoki Y, Hoshida M, Ishikawa M (1995) Comprehensive study on iterative algorithms of multiple sequence alignment. Comput Appl Biosci 11: 13–18

Hogeweg P and Hesper B (1984) The alignment of sets of sequences and the construction of phylogenetic trees. An integrated method. J Mol Evol 20: 175–186

Jimenez-Morales D, Adamian L, Liang J (2008) Detecting remote homologues using scoring matrices calculated from the estimation of amino acid substitution rates of beta-barrel membrane proteins. Conf Proc IEEE Eng Med Biol Soc 2008: 1347–1350

Jennings MJ (1989) Topography of membrane proteins. Annu Rev Biochem 58: 999–1027

Jones DT (1998) Do transmembrane protein superfolds exist? FEBS Lett 423: 281–285

Jones DT (2007) Improving the accuracy of transmembrane protein topology prediction using evolutionary information. Bioinformatics 23: 538–544

Jones DT, Taylor WR, Thornton JM (1994) A mutation matrix for transmembrane proteins. FEBS 339: 269–275

Käll L, Krogh A, Sonnhammer EL (2004) A combined transmembrane topology and signal peptide prediction method. J Mol Biol 338: 1027–1036

Käll L, Krogh A, Sonnhammer EL (2005) An HMM posterior decoder for sequence feature prediction that includes homology information. Bioinformatics 21 (Suppl 1): i251–i257

Karplus K, Barrett C, Hughey R (1998) Hidden Markov models for detecting remote protein homologies. Bioinformatics 14: 846–856

Krogh A, Larsson B, von Heijne G, Sonnhammer EL (2001) Predicting transmembrane protein topology with a hidden Markov model: application to complete genomes. J Mol Biol 305: 567–580

Langosch D and Heringa J (1998) Interaction of transmembrane helices by a knobs-into-holes packing characteristic of soluble coiled coils. Proteins 31: 150–159

Müller T, Rahmann S, Rehmsmeier M (2001) Non-symmetric score matrices and the detection of homologous transmembrane proteins. Bioinformatics 17 (Suppl 1): S182–S189

Möller S, Croning MDR, Apweiler R (2001) Evaluation of methods for the prediction of membrane spanning regions. Bioinformatics 17: 646–653

Ng PC, Henikoff JG, Henikoff S (2000) PHAT: a transmembrane-specific substitution matrix. Predicted hydrophobic and transmembrane. Bioinformatics 16: 760–766

Nugent T and Jones DT (2009) Transmembrane protein topology prediction using support vector machines. BMC Bioinform 10: 159

Pascarella S and Argos P (1992) A data bank merging related protein structures and sequences. Protein Eng 5: 121–137

Pirovano WA (2010) Comparing building blocks of life – sequence alignment and evaluation of predicted structural and functional features. PhD thesis, VU University Amsterdam, ISBN 978-90-8659-419-1

Pirovano W and Heringa J (2008) Multiple sequence alignment. Meth Mol Biol 452: 143–161

Pirovano W, Feenstra KA, Heringa J (2008a) PRALINE™: a strategy for improved multiple alignment of transmembrane proteins. Bioinformatics 24: 492–497

Pirovano W, Feenstra KA, Heringa J (2008b) The meaning of alignment: lessons from structural diversity. BMC Bioinform 23: 556

121

Rost B (1999) Twilight zone of protein sequence alignment. Protein Eng 12: 85–94

Sander C and Schneider R (1991) Database of homology-derived protein structures and the structural meaning of sequence alignment. Proteins 9: 56–68

Schäffer AA, Aravind L, Madden TL, Shavirin S, Spouge JL, Wolf YI, Koonin EV, Altschul SF (2001) Improving the accuracy of PSI-BLAST protein database searches with composition-based statistics and other refinements. Nucleic Acids Res 29: 2994–3005

Shafrir Y and Guy HR (2004) STAM: simple transmembrane alignment method. Bioinformatics 20: 758–769

Simossis VA, Kleinjung J, Heringa J (2005) Homology-extended sequence alignment. Nucleic Acids Res 33: 816–824

Sonnhammer EL, von Heijne G, Krogh A (1998) A hidden Markov model for predicting transmembrane helices in protein sequences. Proc Int Conf Intell Syst Mol Biol 6: 175–182

Stoye J (1998) Multiple sequence alignment with the divide-and-conquer method. Gene 211: GC45–GC56

Thompson JD, Higgins DG, Gibson TJ (1994) CLUSTAL W: improving the sensitivity of progressive multiple sequence alignment through sequence weighting, positions-specific gap penalties and weight matrix choice. Nucleic Acids Res 22: 4673–4680

Tusnády GE, Dosztányi Zs, Simon I (2005) PDB_TM: selection and membrane localization of transmembrane proteins in the protein data bank. Nucleic Acids Res 33: D275–D278

Vogt G, Etzold T, Argos P (1995) An assessment of amino acid exchange matrices in aligning protein sequences: the twilight zone revisited. J Mol Biol 249: 816–831

Wallin E and von Heijne G (1998) Genome-wide analysis of integral membrane proteins from eubacterial, archaean, and eukaryotic organisms. Protein Sci 7: 1029–1038

White SH (2004) The progress of membrane protein structure determination. Protein Sci 13: 1948–1949

White SH and Wimley WC (1998) Hydrophobic interactions of peptides with membrane interfaces. Biochem Biophys Acta 1376: 339–352

Prediction of re-entrant regions and other structural features beyond traditional topology models

Erik Granseth

Department of Genome Oriented Bioinformatics, Technische Universität München, Wissenschaftszentrum Weihenstephan, Freising, Germany

Abstract

A topology model of a membrane protein is a two-dimensional representation of the three-dimensional structure. Most often, it is the only structural information available and it can either come from computer predictions, experiments or a combination of both. However, it has lately become clear that some membrane protein structures contain features that cannot be described by a traditional topology model. They might contain kinks in their transmembrane helices, have interface helices that lie parallel to the membrane surface or contain re-entrant regions that only partially enter the membrane. Since these structural features are almost always functionally important and there are more and more structures available each year, there has been an increasing effort in predicting them. This chapter describes transmembrane helix kinks, interface helices, amphipathic membrane anchors, and re-entrant regions in detail, both from a biological perspective and from the methods that try to predict them. Additionally, prediction of free energy of membrane insertion and Z-coordinates is also covered.

1 Introduction

At first glance, membrane protein topology models seem like a straight-forward concept and good two-dimensional (2D) approximations of three-dimensional (3D) structures. For the majority of known membrane proteins this holds true but as more and more structures become available, some membrane proteins appear

Corresponding author: Erik Granseth, Department of Genome Oriented Bioinformatics, Technische Universität München, Wissenschaftszentrum Weihenstephan, 85350 Freising, Germany (E-mail: erikgr@gmail.com)

too complex to be described by a topology model. This leads to a situation where one has to choose between losing structural information by fitting the structure to a too simple topology model, or perhaps a better choice, redefining the topology concept.

Figure 1 shows one subunit of the homotrimeric sodium-dependent aspartate transporter (Boudker et al. 2007) and is a prime example of a complex membrane protein structure. Apart from having several tilted helices that do not pass through the lipid bilayer perpendicularly, it also has two re-entrant regions (light green and dark green) where the protein chain first enters the membrane, makes a turn, before finally leaving the membrane on the same side as it originated from. In this case, it resembles a helix hairpin that never fully traverses the lipid bilayer.

One other structural feature lost in a standard topology model is the disrupted transmembrane helix. In Fig. 1, the helicity of two transmembrane helices (yellow and orange) is broken inside the membrane.

The structural features described above are unfortunately lost in a traditional topology, but recently more and more topology predictors and other methods have started trying to predict them. There are several reasons for this: basic topology prediction methods perform well, though their prediction accuracies no longer increase as dramatically over existing methods as they once did. As more structural data have become available, it is now possible to develop new methods that were

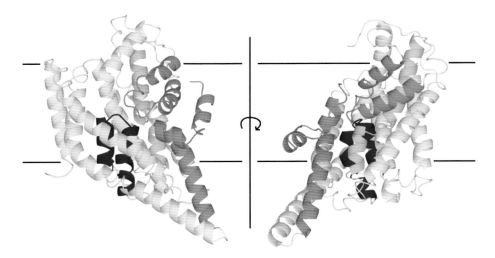

Fig. 1. Structure of the sodium-dependent aspartate transporter (2nwl.pdb). The orange and yellow transmembrane helices are disrupted and the green and dark green show two re-entrant regions. The structure is rotated 180° in the direction of the arrow. The figure was made using PyMOL (DeLano 2002).

impossible perhaps 5 years ago. Finally, features such as re-entrant helices, kinks, and interface helices are almost always functionally important. This chapter will describe such features in greater detail, both from a biological perspective and the methods that try to predict them.

2 Background

2.1 The Z-coordinate as a measure of distance to the membrane

The 3D structures of membrane proteins do not usually contain any information regarding the positioning of the protein within the membrane. The reason for this is that during structure determination, the protein is removed from the membrane and is crystallized using amphiphilic detergent molecules. These are usually disorganized and difficult to assign coordinates. Occasionally, a few lipid molecules and/or detergent molecules may be included in the final structure since they are bound to the crystallized membrane protein, and these can facilitate the determination of the positioning in the membrane.

In order to compare different membrane protein structures to each other they need to have a common coordinate system. To do this, the atom coordinates from the original structure need to be changed so that the structure is positioned in a hypothetical membrane. The simplest way to do this is to calculate a vector that is the average of all the transmembrane helices of the structure and rotate it to the theoretical vector (Wallin et al. 1997). After this, the average hydrophobicity of the amino acids in 1 Å wide slabs is calculated. The coordinate system of the amino acids is then translated so that the hydrophobic maximum is defined as the middle of the membrane ($Z = 0$ Å) and a positive value of the Z-coordinate is directed to the non-cytoplasm and a negative value to the cytoplasm. The Z-coordinate is thus the distance of an amino acid relative to the membrane. More advanced methods have recently been developed, for instance TMDET (Tusnády et al. 2004) and OPM (Lomize et al. 2006) that rotate and translate membrane protein structures to the most likely position inside a membrane.

Figure 2 is an example of a Z-coordinate plot for a rhodopsin membrane protein. It can be seen that the Z-coordinate rises and falls steeply for the transmembrane helices, but also that other structural features such as the C-terminal helix can be clearly recognized in the plot.

3 Interface helices

Close to the membrane–water interface region, membrane proteins sometimes contain helices that run roughly parallel to the membrane surface, see Figs. 2 and 3a.

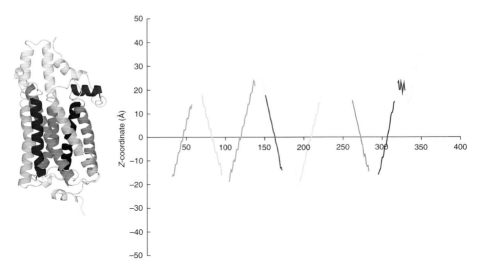

Fig. 2. Structure of rhodopsin and its corresponding Z-coordinate plot of its Cα atoms. The colors of the helices correspond to the colored regions of the Z-coordinate plot.

Such interface helices are located between 15 and 25 Å from the middle of the membrane and have a different amino acid composition than loops and the ends of transmembrane helices in the same region. The most conspicuous difference is that they have almost twice as a high fraction of tryptophan (Trp) and tyrosine (Tyr) residues than loops and transmembrane helices (Granseth et al. 2005). The direction of the aromatic ring of Trp residues is directed toward the center of the membrane when located in an interface helix more than 15 Å from the center of the membrane. This makes it possible to bury the bulky, hydrophobic, six-membered ring inside the hydrophobic bilayer. The interface helices are also less hydrophobic than re-entrant regions and transmembrane helices, but more hydrophobic than average loop regions (Viklund et al. 2006). Interface helices are between 4 and 19 amino acids long with an average of 8.9, almost as long as the average length of a helix in a soluble protein. Interestingly, the more amino acids that are between two adjacent transmembrane helices, the more frequent it is for an interface helix to be present between them. All transmembrane helices separated by more than 31 residues had an interface helix between them. This suggests that interfacial helices might be a way of constraining and positioning transmembrane helices since long loops without interface helices would otherwise have a larger degree of structural freedom.

Not much is known about what functional roles interface helices play although for photosystem I, the interface helices are thought to shield cofactors from the aque-

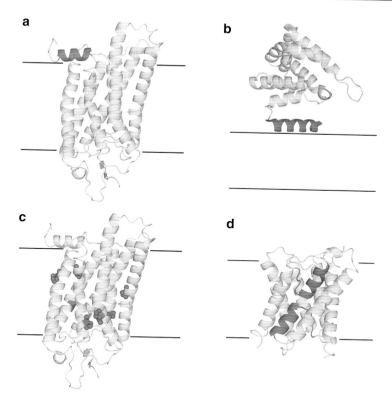

Fig. 3. Examples of structural features that are missed by traditional topology models.
(**a**) Structure of rhodopsin with interface helix colored orange. (**b**) Structure of a monotopic membrane protein that is attached to the membrane with an in-plane membrane anchor (orange). (**c**) Structure of rhodopsin with helical kinks. The orange spheres are the heavy atoms of proline residues and the green is a kink induced by non-proline amino acid, in this case it is a glycine. It is difficult to see the deviation of the helices caused by the kinks because of the viewpoint of the image. (**d**) Structure of aquaporin with its re-entrant regions in orange.

ous phase (Jordan et al. 2001). For the MscS mechanosensitive channel it is thought they are involved in channel gating by transferring mechanical force from a sensor domain to a transmembrane helix in the ion channel (Bass et al. 2003), and for the KirBac 1.1 inward rectifying potassium channel they are thought to be responsible for the regulation of the channel gating (Doyle 2004).

3.1 Prediction of interface helices

The only method to date that has attempted to predict interface helices is TOP-MOD (Viklund et al. 2006). It is an hidden Markov model-based (HMM) method that use sequence profiles provided by BLAST as its input. It uses a rather complex model of an interface helix state compartment since it reflects that residues inside an interface helix can either be directed toward the membrane surface or be directed away from the membrane, a situation that leads to different preferences of amino acids in the two categories. Even though there were more training data for interfacial helices than for re-entrant regions, prediction performance was not very good. Forty-two percent of the interface helices could be detected with 75% specificity using a jack-knifed data-set. This implies that the interface helices are more diverse and the sequence characteristics weak.

3.2 Prediction of amphipathic membrane anchors

Amphipathic membrane anchors are interface helices that attach a non-transmembrane protein to the membrane, see Fig. 3b. This means that the protein does not contain any helices that span the membrane, but is attached to it with a so-called in-plane membrane (IPM) anchor, most often an amphipathic helix located in the interface region, roughly parallel to the membrane. These proteins are called *monotopic* membrane proteins and their amino acid composition in the membrane–water interface region is by and large similar to the interface helices in transmembrane proteins (Granseth et al. 2005). The main differences are that a higher fraction of Trp and Tyr residues of IPM anchors are directed towards the centre of the membrane, while they are composed of fewer acidic amino acids and more basic amino acids compared to the interface helices of transmembrane proteins.

Interestingly, the support vector machine-based method AmphipaSeeK describes that some of its false positive predictions are in fact interface helices of transmembrane proteins (Sapay et al. 2006b). Although the method is trained and developed to predict IPM anchors, there is, at least, partial similarity between them and interface helices. It is also clear that the main problem with predicting IPM anchors is the low sensitivity, around 30% when the specificity is 99.8%. It should be noted that the dataset is small, even by membrane protein standards.

4 Helical kinks in transmembrane helices

Transmembrane helices often contain disruptions of the alpha-helical backbone in the membrane region. These are called helical kinks since they lead to a slight deviation in the direction of the helix (See Fig. 3c), and are frequently responsible for functional diversity in superfamilies (Hall et al. 2009). The most well-known inducer of helical kinks is proline (Pro) due to its unique cyclic side chain that locks its φ backbone dihedral angle at $-75°$. This structural rigidity is what often causes secondary structure disruptions such as membrane helix kinks. However, only about 20%

of all Pro located in transmembrane helices cause helix distortions so a Pro residue is not by itself a clear indicator of whether a kink is present or not (Hall et al. 2009). Yet, Pro is responsible for 35–60% of the distorted membrane helices (Yohannan et al. 2004). Other amino acids that cause kinks are serine, threonine, asparagine, and glutamine.

Transmembrane helix distortions are common; in rhodopsin, 6 out of 7 helices are distorted (Palczewski et al. 2000), and in a larger dataset, 44% out of 405 contained kinks (Hall et al. 2009). Even though Pro is not a perfect kink indicator, much can be learned by looking for conserved Pro in multiple sequence alignments of membrane proteins. Many of the non-Pro kinks found in rhodopsin-like proteins with known three-dimensional structure contained Pro residues at the same positions in their homologous proteins (Yohannan et al. 2004). These positions are called *vestigial prolines* – the hypothesis being that an initial Pro causes a kink, which later becomes locked in the structure by other longer range interactions. After that, the Pro residue serves no particular structural role and can be mutated to a different residue.

4.1 Prediction of helix kinks

The simplest way to predict a helix kink is to create a multiple sequence alignment and look for regions containing conserved Pro residues (Yohannan et al. 2004). When a sequence alignment contained more than 10% Pro, a kink or distortion was present in the corresponding structures, 36 out of 39 times. Another way is to use molecular dynamics (MD) simulations starting from a canonical helix with all the side chains of the amino acids in an extended conformation (Hall et al. 2009). The simulations were able to reproduce the shape of kinked/non-kinked TM helices 70% of the time. If a Pro residue was present in a kinked helix, 79% of the helix simulations were within 1.5 Å of the crystal structure, for the vestigial Pro, 59% were correct but it was possible to increase this by 4–14% by replacing the alternate amino acid with a Pro residue prior to simulation. Finally, only 18% of the non-proline kinks were reproduced.

5 Re-entrant regions

A re-entrant region is a membrane penetrating part of the protein that enters the membrane and then exits again from the same side, see Fig. 3d. There is no clear cut definition of what constitutes a re-entrant region, but one sensible way of defining them is basing it on the Z-coordinate of the residues. Different studies have used different ways of defining what a re-entrant region is, which makes direct comparisons of results somewhat ambiguous. Viklund et al. first defined a re-entrant region as a part of the sequence that penetrates the membrane between 3 and 25 Å and start and end on the same side of a membrane located at 15 Å. Regions penetrating between 1.5 and 3 Å were also included if they monotonically decreased/increased with regard to the deepest lying residue. Re-entrant regions were also subdivided into three more categories depending on their secondary structure. A helix hairpin enters and exits the membrane as helices which are connected to each other by a loop (helix–coil–helix, see Fig. 1). The second category contains regions that have one helix followed by a coil or coil followed by helix (helix–coil/coil–helix) and for instance includes the re-entrant region of aquaporin (Fig. 3d). The third and final category is a re-entrant region that only has irregular secondary structure (coil). These definitions were used to train the first machine learning methods to detect these structures (Viklund et al. 2006).

Later methods changed the way of annotating re-entrant regions to be classified only by the Z-coordinates of the amino acids and not by the secondary structure. Nugent and Jones (2009) assigned a part of the sequence that penetrates at least 6 Å but not more than 6 Å from the opposite membrane face as a re-entrant region. A recent study by Viklund and Elofsson (2008) subdivided annotations into three different classes, a membrane dip (roughly equivalent to the coil described above), a re-entrant region (equivalent to the Nugent et al. definition) and a TM hairpin which is basically two short membrane helices that penetrate further than 6 Å from the opposite membrane border but never crosses it.

Not so much is known about the sequence characteristics of re-entrant regions, but they are on average less hydrophobic than transmembrane helices but more so than interface helices (Viklund et al. 2006). They are also commonly enriched with small amino acids such as glycine and alanine. This is not surprising since glycine is frequently found in coil regions that change the direction of the amino backbone by 180°, which all re-entrant regions do. The deepest penetrating regions also contain more helix secondary structure than more shallow ones.

5.1 Prediction of re-entrant regions

5.1.1 TOP-MOD

TOP-MOD is an HMM-based method, which has demonstrated a sensitivity of 0.69 and specificity of 0.72 when detecting re-entrant regions. These results are however based on known transmembrane helix locations, which is a somewhat artificial situation since re-entrant regions might be mispredicted as transmembrane helices and vice versa. Under a more realistic scenario, where both re-entrant- and transmembrane-helices are predicted, the sensitivity drops to 0.47 with the same specificity. The main source of the decrease in sensitivity is that 8 out of 36 re-entrant regions were falsely predicted as transmembrane helices, coming mainly from the longer helix–coil–helix category.

5.1.2 TMloop

TMloop is not based on any machine learning algorithm, instead it uses collective motifs to identify re-entrant or membrane dipping loop regions (Lasso et al. 2006). The re-entrant regions were identified in structurally determined membrane proteins and sequence alignments were created to identify sequence motifs. In order to avoid false positive matches, the motifs were evaluated against known non-re-entrant regions in a large number of membrane proteins. The method uses three different types of pattern matching: identical motifs, motifs-based on chemical equivalency, and finally, motifs using structural equivalency. It is also possible to either use a single motif mode or a collective motif mode where the latter mode permits non-exact matching and uses a set of partially overlapping patterns. This makes it possible to find more distantly related re-entrant regions.

The main drawback of a motif-based method is that it is only possible to identify re-entrant regions that are already known and characterized.

5.1.3 OCTOPUS

The first method that fully integrated a model for re-entrant regions into its topological grammar was OCTOPUS (Viklund and Elofsson 2008). It uses a combination of hidden Markov models and artificial neural networks for its predictions. Conceptually, it is similar to combining Z-coordinate predictions with a hidden Markov model for the final topology prediction. Four neural networks predict which region an amino acid belongs to based on their position-specific substitution matrix provided by a BLAST search. The networks are predicting: membrane (± 0–$13\,\text{Å}$), interface (± 11–$18\,\text{Å}$), loop (± 13–$23\,\text{Å}$), and globular (± 23–$\,\text{Å}$), amino acids. The output from these network combined with additional neural networks trained on other sequence characteristics are then fed into a hidden Markov model that in addition to predicting transmembrane helices also is able to predict re-entrant helices/membrane dipping loops and helix–helix hairpins (See Fig. 1). It is able to find 10 of 49 re-entrant helices/membrane dipping loops and only two false positive predictions in a cross-validated dataset. Four out of seven hairpins were correctly predicted with no false positives. Combined with its 94% accuracy in predicting the correct topology, it is clearly one of the most capable methods available today.

5.1.4 MEMSAT-SVM

One other method that also performs really well is MEMSAT-SVM, a support vector machine-based method that also predicts re-entrant regions (Nugent and Jones

2009). It achieves 89% accuracy in predicting the correct topology of its own cross-validated dataset, on which OCTOPUS only has 79% accuracy. Seven of eleven proteins with re-entrant regions were correctly predicted by MEMSAT-SVM, on which OCTOPUS was correct on eight. It seems as if MEMSAT-SVM is better at predicting the topology of "normal" membrane proteins, whereas OCTOPUS is slightly better at predicting re-entrant regions. It is, however, difficult to assess until independent benchmarks are performed.

6 Prediction of the Z-coordinate

ZPRED is a novel method that tries to predict how deep inside the membrane an amino acid is located (Granseth et al. 2006). In order to do this, it uses artificial neural networks trained on a combination of sequence profiles and topology predictions to estimate the position of the amino acids in the 5–25 Å region. It is important to note that ZPRED does not provide any information about the orientation of the protein since it is trained on the absolute value of the Z-coordinates, see Fig. 4. The predictions are also constrained to treat all residues more distant than 25 Å from the membrane center as a single "globular" region and residues closer than

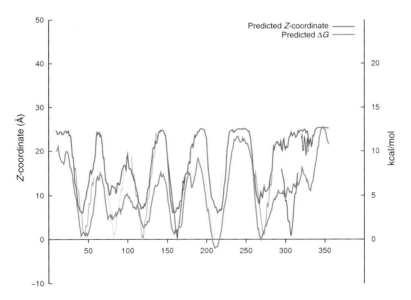

Fig. 4. Example of a Z-coordinate and ΔG prediction of the same rhodopsin molecule depicted in Fig. 1. Note that the light gray plot is the absolute value of the Z-coordinate of the C$^\alpha$ atoms since that is what ZPRED (red) is trained to predict. The ΔG prediction is in green.

5 Å to the center as a single "membrane core" region. The reasoning behind this is that the environment in these regions is rather homogeneous, either hydrophilic or hydrophobic, and instead focuses predictions to the regions where the environment changes the most. The average error of ZPRED's prediction is 2.55 Å and two-thirds of the amino acids are predicted within 3 Å of the correct Z-coordinates in a cross-validated test.

ZPRED was later improved by altering window sizes, inclusion of polarity based encoding, post-processing etc. and incorporation of two different networks, one for single TM proteins and one for multi-helix membrane proteins (Papaloukas et al. 2008). Since this lead to a large amount of parameters to optimize with only a relatively small amount of training data, principal component analysis was used to decrease the dimension of the input data from 551 to 13, which described 99% of the variation of the original data. ZPRED2 was able to decrease the average error to 2.18 Å.

7 Free energy of membrane insertion ΔG

One method that is worth mentioning even though it does not predict re-entrant regions or interface helices is SCAMPI, since it uses a novel way of predicting membrane protein topology (Bernsel et al. 2008). Instead of training a machine learning method on known membrane protein sequences/profiles, it is based on experimental observations of membrane helix insertion into the lipid bilayer by the translocon (Hessa et al. 2005, 2007). The position-dependent free energy of membrane insertion (ΔG_{app}) was measured for each of the 20 amino acids using a model system consisting of larger protein attached to a helix consisting of alanine and leucine residues. The helix was then altered by having one or several of the same kind of amino acid at different positions in the helix. The ΔG_{app} could then be determined by measuring the fraction of helices that was inserted into the membrane versus the fraction that was not inserted. The experimental results were then modeled by single or double Gaussian functions. The total ΔG_{app} for a natural or synthetic transmembrane helix could then be described by an equation, see Eq. (1), and an example prediction is shown in Fig. 4.

$$\Delta G_{app} = \sum_{i=1}^{l} \Delta G_{app}^{aa(i)} + 0.27 \sqrt{\left(\sum_{i=1}^{l} \Delta G_{app}^{aa(i)} \cdot \sin(100° \cdot i) \right)^2 + \left(\sum_{i=1}^{l} \Delta G_{app}^{aa(i)} \cdot \cos(100° \cdot i) \right)^2}$$
$$+ 9.3 - 0.65l + 0.0082l^2 \tag{1}$$

In Eq. (1) the first term sums up the individual contributions of each amino acid. The next part is a cyclical term describing the hydrophobic moment since the contributions sometimes differ when two amino acids are facing the same side of the helix. ΔG_{app} is also dependent on the length of the helix.

To summarize, instead of being based on statistics derived from membrane proteins with known topology, data from physical experiments were used.

Two different predictions methods were created: one that simply uses a sliding-window scanning the protein sequence calculating ΔG_{app} values. In the resulting curve, minima are assumed to be the transmembrane helices and the positive-inside rule (von Heijne 1992) is used to orient the prediction. The second method, SCAMPI, is model-based. SCAMPI is similar to hidden Markov models, but without transition probabilities. It has four different modules, the membrane compartment (M), inside loop (I), outside loop (O), and inside loop close to the membrane (i). The reason for the last module is to incorporate the positive-inside rule due to the overrepresentation of arginines and lysines in inside loops. The membrane module uses Eq. (1) to calculate the free energy of insertion and converts it to the corresponding estimated insertion probability used as an emission probability, whereas the i, I, and O modules use flat distributions for amino acid emission probabilities. The two methods are also capable of using homology information to yield more accurate predictions.

Both methods are reported to have similar performance, around 80% correctly predicted topologies, similar to MEMSAT3 and PRODIV-TMHMM. It is however slightly worse at discriminating globular proteins from membrane proteins than other methods.

8 The frequency of re-entrant regions and interface helices

As more membrane protein topology prediction methods are developed and more genomes are sequenced, the general consensus appears to be that around 25–30% of the encoded proteins in a genome are membrane proteins. How many re-entrant regions do these contain? Among the known membrane protein structures, around 5% contain re-entrant loops (corresponding to the helix–coil–helix and coil/helix classes described above; Cuthbertson et al. 2005). These proteins are from various organisms and do not say anything about how frequent the regions are in a genome. Predictions using TOP-MOD estimate that around 10–15% of the membrane proteins of a genome contain re-entrant regions (Viklund et al. 2006). However, MEMSAT-SVM predicts a much smaller fraction, around 2–3% (Nugent and Jones 2009). The main reason for this significant difference is probably the different definitions of what a re-entrant region is.

Table 1. Methods and web-servers that predict structural features of membrane protein beyond the traditional topology

Method	Focus	URL	References
Topcons	Re-entrant, Z-coordinate, ΔG	topcons.cbr.su.se	Bernsel et al. (2009)
TMloop	Re-entrant	membraneproteins.swan.ac.uk/TMloop	Lasso et al. (2006)
TOP-MOD	Re-entrant, interface helix		Viklund et al. (2006)
MEMSAT-SVM	Re-entrant	bioinf.cs.ucl.ac.uk/psipred/	Nugent and Jones (2009)
Kinks	Kinks		Hall et al. (2009)
AmphipaSeeK	Amphipathic membrane anchors	pbil.ibcp.fr/htm/pbil_ibcp_Amphipaseek.html	Sapay et al. (2006a)

Topcons is a meta-prediction method that integrates the results from several different methods (ZPRED, SCAMPI, OCTOPUS, and PRODIV).

9 Summary

- The traditional topology predictions methods of today are good at detecting which regions that are located inside the membrane.
- Many membrane proteins contain structural features that are missed by traditional topology models and predictions methods, for instance interface helices, re-entrant regions, and kinks in transmembrane helices.
- Interface helices are parallel to the membrane and located in the membrane–water interface region.
- Re-entrant regions enter the membrane, then exit again from the same side. They never completely cross the membrane.
- Kinks break the helicity of transmembrane helices and are often caused by proline residues.
- Several new topology prediction methods try to predict these features, but the performance is not on par with the performance of transmembrane helix detection.
- Other new prediction methods try to predict physical features such as the free energy of insertion of a transmembrane helix or how deep into the membrane an amino acid is located.
- The references and URLs of the different methods are summarized in Table 1.

References

Bass RB, Locher KP, Borths E, Poon Y, Strop P, Lee A, Rees DC (2003) The structures of BtuCD and MscS and their implications for transporter and channel function. FEBS Lett 555: 111–115
Bernsel A, Viklund H, Falk J, Lindahl E, von Heijne G, Elofsson A (2008) Prediction of membrane–protein topology from first principles. Proc Natl Acad Sci USA 105: 7177–7181
Bernsel A, Viklund H, Hennerdal A, Elofsson A (2009) TOPCONS: consensus prediction of membrane protein topology. Nucleic Acids Res 37: W465–W468
Boudker O, Ryan RM, Yernool D, Shimamoto K, Gouaux E (2007) Coupling substrate and ion binding to extracellular gate of a sodium-dependent aspartate transporter. Nature 445: 387–393

Cuthbertson JM, Doyle DA, Sansom MS (2005) Transmembrane helix prediction: a comparative evaluation and analysis. Protein Eng Des Sel 18: 295–308

DeLano WL (2002) The PyMOL molecular graphics system. DeLano Scientific, Palo Alto, USA

Doyle DA (2004) Structural themes in ion channels. Eur Biophys J 33: 175–179

Papaloukas C, Granseth E, Viklund H, Elofsson A (2008) Estimating the length of transmembrane helices using Z-coordinate predictions. Pro Sci 17: 271–278

Granseth E, von Heijne G, Elofsson A (2005) A study of the membrane–water interface region of membrane proteins. J Mol Biol 346: 377–385

Granseth E, Viklund H, Elofsson A (2006) ZPRED: predicting the distance to the membrane center for residues in alpha-helical membrane proteins. Bioinformatics 22: E191–E196

Hall SE, Roberts K, Vaidehi N (2009) Position of helical kinks in membrane protein crystal structures and the accuracy of computational prediction. J Mol Graph Model 27: 944–950

Hessa T, Kim H, Bihlmaier K, Lundin C, Boekel J, Andersson H, Nilsson I, White SH, von Heijne G (2005) Recognition of transmembrane helices by the endoplasmic reticulum translocon. Nature 433: 377–381

Hessa T, Meindl-Beinker NM, Bernsel A, Kim H, Sato Y, Lerch-Bader M, Nilsson I, White SH, von Heijne G (2007) Molecular code for transmembrane-helix recognition by the Sec61 translocon. Nature 450: 1026–1030

Jordan P, Fromme P, Witt HT, Klukas O, Saenger W, Krauss N (2001) Three-dimensional structure of cyanobacterial photosystem I at 2.5 A resolution. Nature 411: 909–917

Lasso G, Antoniw JF, Mullins JG (2006) A combinatorial pattern discovery approach for the prediction of membrane dipping (re-entrant) loops. Bioinformatics 22: e290–e297

Lomize AL, Pogozheva ID, Lomize MA, Mosberg HI (2006) Positioning of proteins in membranes: a computational approach. Protein Sci 15: 1318–1333

Nugent T and Jones DT (2009) Transmembrane protein topology prediction using support vector machines. BMC Bioinform 10: 159

Palczewski K, Kumasaka T, Hori T, Behnke CA, Motoshima H, Fox BA, Le Trong I, Teller DC, Okada T, Stenkamp RE, Yamamoto M, Miyano M (2000) Crystal structure of rhodopsin: a G protein-coupled receptor. Science 289: 739–745

Sapay N, Guermeur Y, Deleage G (2006a) Prediction of amphipathic in-plane membrane anchors in monotopic proteins using a SVM classifier. BMC Bioinform 7: 255

Sapay N, Montserret R, Chipot C, Brass V, Moradpour D, Deleage G, Penin F (2006b) NMR structure and molecular dynamics of the in-plane membrane anchor of nonstructural protein 5 A from bovine viral diarrhea virus. Biochemistry 45: 2221–2233

Tusnády GE, Dosztányi Z, Simon I (2004) Transmembrane proteins in the protein data bank: identification and classification. Bioinformatics 20: 2964–2972

Viklund H and Elofsson A (2008) OCTOPUS: improving topology prediction by two-track ANN-based preference scores and an extended topological grammar. Bioinformatics 24: 1662–1668

Viklund H, Granseth E, Elofsson A (2006) Structural classification and prediction of reentrant regions in alpha-helical transmembrane proteins: application to complete genomes. J Mol Biol 361: 591–603

von Heijne G (1992) Membrane protein structure prediction. Hydrophobicity analysis and the positive-inside rule. J Mol Biol 225: 487–494

Wallin E, Tsukihara T, Yoshikawa S, von Heijne G, Elofsson A (1997) Architecture of helix bundle membrane proteins: an analysis of cytochrome c oxidase from bovine mitochondria. Protein Sci 6: 808–815

Yohannan S, Faham S, Yang D, Whitelegge JP, Bowie JU (2004) The evolution of transmembrane helix kinks and the structural diversity of G protein-coupled receptors. Proc Natl Acad Sci USA 101: 959–963

Dual-topology: one sequence, two topologies

Erik Granseth

Department of Genome Oriented Bioinformatics, Technische Universität München, Wissenschaftszentrum Weihenstephan, Freising, Germany

Abstract

The function of a membrane protein is dependent on that it is inserted into the lipid bilayer in a correct way. Intriguingly, for a small number of membrane proteins, there is growing evidence that they have flexible topologies. Some of them have a varying number of helices inserted into the membrane, or have the same number of transmembrane helices, but different membrane spanning regions. Others are inserted with opposite topologies in the membrane in an approximate 1:1 ratio, forming antiparallel homodimers. Thus, the same sequence can code for more than one topology. During the last few years, there have been increasing efforts in studying topologically flexible proteins since they might hold clues about the evolution of membrane proteins.

1 Introduction

Most membrane proteins are topologically stable and are always inserted into the membrane in a unique way since the directionality of transport across the membrane is well defined, and an incorrect topology would disrupt the function of the protein. There is, however, growing evidence that a small number of membrane proteins have flexible topologies and during the past few years there have been increasing efforts in studying these, both out of curiosity and because they might hold clues about the evolution of membrane proteins.

Figure 1 shows an example of a membrane protein, EmrE, which exhibits dual-topology in its crystal form (Chen et al. 2007). The structure forms a homodimer where the subunits exhibit opposite topologies while sharing the same amino acid

Corresponding author: Erik Granseth, Department of Genome Oriented Bioinformatics, Technische Universität München, Wissenschaftszentrum Weihenstephan, 85350 Freising, Germany (E-mail: erikgr@gmail.com)

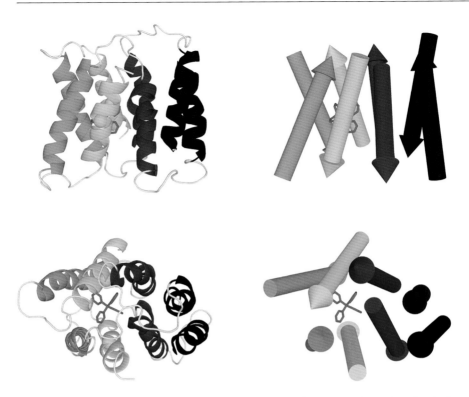

Fig. 1.   Structure of EmrE from *E. coli* in a dual-topology conformation where the subunits have opposite topologies. The coloring of the transmembrane helices is the same for both subunits (transmembrane helix 1 is yellow, 2 light green, 3 dark green, and 4 almost black). The gray molecule in the middle is the bound substrate. The top part of the picture shows the structure viewed from the side of the membrane and the lower part shows it from top. The right parts show the TMH as idealized arrows for clarity. The figure was generated using PyMOL (DeLano 2002).

sequence. This demonstrates that some protein sequences can be inserted either way into the membrane, thus having dual-topology. The topology of EmrE is a controversial topic; some reports show that it cannot have dual-topology in its functional state (Soskine et al. 2006) while another shows that it can (Rapp et al. 2007). Other membrane proteins with potential dual-topology are also controversial, but it is difficult to settle the arguments since it is possible that experimental techniques used to evaluate the topology and function might actually alter them (Soskine et al. 2006). Fortunately, comparative genomics makes it possible to draw some conclusions about the existence of dual-topology and the evolution of membrane proteins.

This chapter summarizes what is currently known about dual-topology proteins, discusses what the implications are with regard to membrane protein evolution and how dual-topology proteins can be found and examined by bioinformatic methods.

2 Background

2.1 A brief history of dual-topology research

The word dual-topology was first mentioned in 1989 by Parks et al. (1989) in experiments with viral hybrid genes. They discovered that by fusing the N-terminus of the M$_2$ polypeptide from *influenza A* with the signal anchor and ectodomain from *paramyxovirus* protein HN, the resulting hybrid molecule was inserted into the membrane in two opposite orientations. In 1994, two groups reported that the large envelope protein (protein L) of *Hepatitis B* virus appeared to have two different topologies, one with three transmembrane helices when located inside the endoplasmic reticulum (ER) membrane and the one with either three or four when located in the virion envelope, see Fig. 3b (Bruss et al. 1994; Ostapchuk et al. 1994). However, they did not comment on dual-topology until a year later when they continued their research on the properties of the *Hepatitis B* envelope proteins (Bruss and Vieluf 1995; Prange and Streeck 1995). Since these first reports of atypical behavior of some viral membrane proteins, dual-topology proteins have now been found in all three domains of life.

2.2 The difference between dual- and multiple-topology

Dual-topology refers to membrane proteins that are inserted with opposite topologies in the lipid bilayer, in an approximate 1:1 stoichiometry (von Heijne 2006). They are usually small, around 100 amino acids, do not have very long loops between their transmembrane helices, and have a small number of arginines and lysines (Rapp et al. 2006). Figure 1 shows an example of a dual-topology membrane protein.

Multiple-topology refers to when a membrane protein either has varying number of helices inserted into the membrane, or has the same number of transmembrane helices, but different regions are membrane inserted. Multiple topologies are caused by one or more marginally hydrophobic and therefore inefficiently inserted transmembrane helices (von Heijne 2006).

These definitions will be used throughout the rest of this text, but in others multiple-topology is often denoted dual-topology.

2.3 Topology mapping

In order to study the topologic behavior of a membrane protein, it is necessary to experimentally verify its topology. The most common way of doing this is by topology mapping, where the location of specific amino acids is determined – whether they are located in the cytoplasm, inside the membrane, or in the extra-cytoplasmic space. Antibodies or biochemical agents can be used to do this (Kimura et al. 1997), but another common technique is to fuse a foreign protein domain to a loop region.

The domain is only active when it is located on one side of the membrane and inactive on the opposing side (Manoil 1991). The methodology has been used to study the topologic behavior of EmrE by fusing green fluorescent protein (GFP) or alkaline phosphatase (PhoA) to its C-terminus (Rapp et al. 2006). GFP is the only fluorescent when it is located inside the cytoplasm whereas PhoA is only functional when it is located in the periplasm. This makes it possible to study how sensitive the topologies of dual-topology candidates are to amino acid mutations. One potential drawback with the protein domain fusion technique is that the fused domains are often large and little is known about their influence over topology.

2.4 Arginines and lysines are important for the topology

The best known sequence characteristic of a membrane protein, apart from the hydrophobic transmembrane helices, is the "positive inside" rule. It states that the majority of the positively charged amino acids, arginines (R) and lysines (K), in the loops are located on the cytoplasmic side of the membrane. There is an almost twice as high a fraction of K and R on the cytoplasmic side compared to the extra-cytoplamic side (Granseth et al. 2005) and genome-wide studies have confirmed that the bias is present in almost all organisms (Wallin and von Heijne 1998; Nilsson et al. 2005).

Dual-topology proteins generally have few positively charged amino acids in their loops and the difference in their numbers between the two sides, the KR-bias, is close to zero, leading to the possibility of being inserted either way into the membrane. Topologically stable membrane proteins usually have a KR-bias >2 or <–2.

2.5 Internal structural repeats – evidence of former gene duplication events

One of the unexpected results that became apparent when more and more membrane protein structures were determined was that up to 50% of them had a clear, internal symmetry which was particularly common among transporters (Choi et al. 2008). The symmetry is evident when the structure is divided into two or more parts and aligning them to each other, resulting in a root mean square deviation of 2–4 Å. It is most common with a twofold internal symmetry, for example, the aquaporin structure, but there are also cases of threefold symmetry, such as in cytochrome c oxidase. The symmetry can be explained as evidence of an ancient gene duplication followed by a gene-merging event. This is the most common way of topology evolution and usually leads to a duplication of the number of transmembrane helices, although it is also possible with only partial duplications (Shimizu et al. 2004).

Usually it is impossible to detect the internal symmetry by sequence information alone, since sequence identities range between 10% and 25%, and the components have simply diverged too far from each other. However, by combining homology information with careful Smith-Waterman alignments, Choi et al. (2008) predicted that around 25% of the membrane proteins in Swiss-Prot had probable internal repeat symmetry.

Twofold internal symmetry can be divided into two separate classes: a membrane protein where the two halves have an even number of transmembrane helices will have a parallel internal symmetry with a symmetry axis perpendicular to the membrane plane, see right part of Fig. 2. A membrane protein where the two halves have an odd number of transmembrane helices will have an antiparallel symmetry, with its axis of symmetry in line with the plane of the membrane. This is a particu-

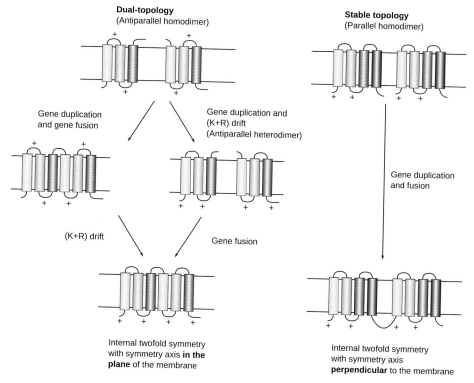

Fig. 2. **Possible evolutionary steps for membrane proteins with internal symmetry.** A dual-topology protein with an even number of transmembrane helices would need a de novo creation or a loss of a membrane helix in order for a possible merge to occur after the duplication.

larly intriguing situation when taking the evolutionary explanation behind internal symmetry into consideration since half of the merged protein would contradict the "positive-inside" rule. The reason why dual-topology proteins attract so much attention is that they have a KR-bias close to zero and are therefore suitable as precursors for membrane proteins with internal antiparallel structural symmetry, see left part of Fig. 2.

3 Prediction of dual-topology

There are no easy and straightforward ways of predicting whether a protein has dual-topology or not. Currently, the only successful approach has been the use of comparative genomics in combination with sequence characteristics (Rapp et al. 2006). Pfam was used to scan 174 fully sequenced prokaryotic genomes and classify each protein into one or several Pfam families (Finn et al. 2008). The topology of each member of the families was predicted by TMHMM (Krogh et al. 2001) and the sequences of the families were aligned by ClustalW (Thompson et al. 1994). This made it possible to create "consensus topologies" where each position in the multiple sequence alignment was assigned as residing within the membrane or belonging to a loop. This was achieved by choosing the most frequent state and filtering so that the consensus membrane helices were sufficiently long. The KR-bias was then calculated for each sequence in the family by adding lysine and arginine residues located in odd loops and subtracting them when found in even loops. The consensus topology was necessary to decrease mispredictions that would have occurred had the individual TMHMM predictions of each sequence been used, since a missing transmembrane helix would have led to a drastically different KR-bias.

What emerged when looking at the location of the genes that encoded the known dual-topology proteins was that they occurred either as closely spaced pairs or as singletons in the genomes. The closely spaced pairs encoded homologous proteins with opposite topologies whereas singletons encoded proteins with a KR-bias close to zero, implying that they had dual-topology. Of the Pfam families searched in the 174 genomes, only the small multidrug resistant (SMR) family, the CrcB family (involved in camphor resistance) and DUF606 (a family with unknown function) seemed to have dual-topology members and possibly also the GlpM and UPF600 families.

3.1 The small multidrug resistance family: one family, different topologies

One of the most intriguing and controversial membrane protein families is the SMR. Members of this family export a wide range of toxins and polyaromatic cations from the cell by coupling transport to the influx of protons (Chung and Saier 2001; Schuldiner et al. 2001). The SMR proteins exist mainly in bacteria but also in

142

some archaeal and eukaryotic organisms. They are approximately 110 amino acids long and have four transmembrane helices, which make them compact, without any long, protruding loops that extend from the membrane.

Two of its members in *Escherichia coli*, EmrE (shown in Fig. 1) and SugE, most likely have dual-topology but interestingly, two other members, YdgE and YdgF do not have dual-topology and have opposite topology to each other (Rapp et al. 2006). The *ydgE* and *ydgF* genes overlap on the *E. coli* chromosome and their proteins only catalyze drug efflux when they are coexpressed (Nishino and Yamaguchi 2001). The two proteins are therefore thought to form an antiparallel heterodimer in the lipid bilayer. EmrE and SugE each correspond to the top left part of Fig. 2 and YdgE/F the middle part, after the gene duplication and K + R drift. The evolutionary hypothesis is that a singleton gene encoding a dual-topology protein undergoes gene duplication. After that, the genes mutate and evolve opposite KR-biases so that the two resulting proteins become fixed in opposite orientations. The two genes might also fuse into a single gene, see Fig. 2, but this has not been found in the members of the SMR family, possibly because it would require the de novo creation of a transmembrane helix. However, it has been found in other families, see the DUF606 family below.

EmrE has been extensively studied and mutated to test the proposed evolutionary hypothesis. One especially illuminating experiment was to emulate the evolution by converting one gene encoding a homodimeric EmrE to two genes encoding heterodimeric EmrE, where each subunit had opposite topology to the other (Rapp et al. 2007). It took three mutations to create a N_{in}–C_{in} version and three more to create a N_{out}–C_{out} version. When expressing only the N_{in}–C_{in} version of EmrE or N_{out}–C_{out} version individually, the *E. coli* cells were unable to survive in high concentrations of a toxic compound, ethidium bromide, whereas the cells survived when the proteins were coexpressed, strongly suggesting a functional, antiparallel heterodimer. This supports the notion that dual-topology proteins are progenitors of membrane proteins with antiparallel internal symmetry.

3.2 The DUF606 family contains fused genes

Most members of the DUF606 family contain five transmembrane helices but there are also members that are twice as long and have 9 or 10 helices (Rapp et al. 2006). For the larger proteins, the first and second halves are homologous to each other, with around 20–35% sequence identity, and they have opposite KR-bias. Proteins with five transmembrane helices had a KR-bias close to zero when found as singletons and had opposite KR-biases when found in closely spaced pairs. Thus the DUF606 family seem to be the first family found containing members in all the different stages of the evolutionary hypothesis.

143

4 Examples of membrane proteins with dual- or multiple-topology

There are of course other examples of membrane proteins with dual- or multiple-topology and here follows a few of them. There are also some examples that use topology as a targeting system.

4.1 MRAP

The melanocortin-2 (MC2) receptor accessory protein (MRAP) contains a single transmembrane helix and is involved in trafficking of the G-protein-coupled MC2 receptor to the plasma membrane (Sebag and Hinkle 2007). The MC2 receptor remains non-glycosylated or core glycosylated without the presence of MRAP and this leads to retention in the endoplasmic reticulum (ER) and subsequent degradation (Petaja-Repo et al. 2001). Both antibodies that bind to the N- and C-terminal tails and glycosylation studies indicate that MRAP exists in both N_{out}–C_{in} and N_{in}–C_{out} orientations and they form an antiparallel homodimer in the membrane, see Fig. 3a. Molecular complementation studies with fragments of yellow fluorescent proteins attached to the termini of MRAP revealed that dual-topology was present in both the ER and the plasma membrane (Sebag and Hinkle 2009).

4.2 Ductin

Ductin is similar to the F_0 subunit of F_1F_0 ATPase and is predicted to have four transmembrane helices (Dunlop et al. 1995). Six ductin molecules either make up a functional unit for the core structure of V_0 (a part of a vacuolar H^+-ATPase, V-ATPase), or a component of a connexon channel in gap junctions (the structure that connects neighboring cells). When ductin is a part of V-ATPase, it has a N_{out}–C_{out} topology (where the termini are facing the luminal side of the ER) and when part of a connexon channel, it has a N_{in}–C_{in} topology, where the termini are inside the cytoplasm of the cell. Dunlop et al. show in in vitro studies that ductin is co-translationally inserted in dual orientations in microsomal membranes in an approximate 1:1 ratio thus exhibiting dual-topology.

Interestingly, alterations of conserved charged residues in the N-terminal region did not change the ratio of the two opposing forms of the protein (Dunlop et al. 1995). The replacement of acidic amino acids with basic ones had no effect on the ratio of the two conformations. This is surprising since the topology of other dual-topology proteins often change into one or the other orientation when the KR-bias is altered.

4.3 *Hepatitis B* virus L protein

One single open reading frame of *Hepatitis B* virus encodes three homologous membrane proteins: small (S), medium (M), and large (L) envelope proteins. Since the three proteins have different start codons but share open reading frame, the 226 ami-

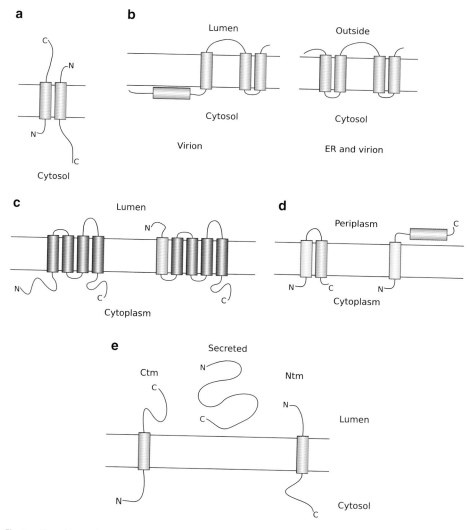

Fig. 3. **Topology of potential dual-topology membrane proteins.** (**a**) Dual-topology of the melanocortin-2 receptor accessory protein. The non-cytoplasmic region in the top part of the figure can either be the lumen of the endoplasmic reticulum or be the periplasm. (**b**) Multiple topologies of *Hepatitis B* virus L protein. (**c**) Multiple topologies of the *Hepatitis C* virus protein NS4B. (**d**) Multiple topologies of twin-arginine motif translocase subunit A, TatA. (**e**) Multiple topologies of the prion protein, PrP.

no acids of S are present in both M and L, and M is present in L. The proteins have four transmembrane helices and interestingly, the L protein has two different topologies. During maturation, approximately half of the L proteins post-translationally translocate their N-terminal region across the ER membrane. The multiple-topology is preserved in the viral envelope after the virus buds from its host cell and by retaining their N-terminal domain inside and outside the viral capsid, the L protein serves dual functions of capsid envelopment and receptor binding (Lambert and Prange 2001). The two different topologies differ in that one has three membrane helices and the other has four, the N-terminal helix is either inside the membrane or not, see Fig. 3b (Lambert et al. 2004).

4.4 *Hepatitis C* virus protein NS4B

The non-structural protein 4B from *Hepatitis C* virus causes membrane changes of the ER leading to the formation of a membranous web, which is essential for the replication of the virus (Appel et al. 2005). NS4B has four transmembrane helices immediately after translation with the N- and C-termini located in the cytosol. After processing, some of the molecules have an additional fifth N-terminal transmembrane helix, with the N-terminal end in the ER lumen, see Fig. 3c. This location of transmembrane helix is suggested to be associated with the membrane-changing capacity of NS4B. The multiple-topology of NS4B is also a conserved feature occurring in all seven of the genotypes included in the study (Lundin et al. 2006).

4.5 TatA

The Tat system is able to transport folded proteins through the cytoplasmic membrane or the thylakoid membrane (Weiner et al. 1998). The proteins that the Tat system transports have a "twin-arginine" motif in their N-terminal signal sequences and hence the name. TatA is considered to be the pore forming subunit and forms homo-oligomeric complexes in the membrane, where each TatA molecule contains one N-terminal transmembrane helix followed by an amphiphilic helix, and a hydrophilic C-terminus. The N-terminus was first reported to be located in the periplasm (Gouffi et al. 2004), but later studies revealed that it resides in the cytoplasm (Chan et al. 2007) whereas the amphiphilic helix is located either in the periplasm or inside the membrane as a second transmembrane helix, see Fig. 3d. The two different conformations are dependent on membrane potential. There is also some speculation that the pore adjusts the number of TatA molecules depending on the size of the substrate (Oates et al. 2005).

4.6 PrP

The prion protein (PrP) is considered to be the infectious agent behind diseases like Creutzfeld-Jakob in humans, mad cow disease in cattle and scrapie in sheep. The hypothesis is that PrP exists in two different conformations, PrP^C, which is a normal cell-surface glycoprotein and PrP^{Sc}, an altered isoform which can convert PrP^C into more PrP^{Sc} which then eventually causes spongiform destruction of brain tissue (Prusiner 1997). The underlying biochemical processes behind this are not completely understood, but some argue that there exists another form of PrP, ^{Ctm}PrP, that is attached to the membrane by a transmembrane helix whereas PrP^C is secreted and attached to the ER membrane by a GPI-anchor in its C-terminus. Interestingly, there is one more form, ^{Ntm}PrP, which has inverted topology compared to ^{Ctm}PrP, see Fig. 3e (Hegde et al. 1998a). There is speculation that a proportion of the variants rely on unidentified accessory proteins that interacts with translocation apparatus in the ER (Hegde et al. 1998b).

^{Ctm}PrP also has an unusual dual mode of attachment to the membrane – it has both a transmembrane helix and a C-terminal glycolipid anchor. Some evidence indicates that ^{Ctm}PrP is involved in neurodegenerative diseases since known disease-causing mutations were demonstrated to lead to an increase of ^{Ctm}PrP in vitro (Hegde et al. 1999). However, later studies revealed that only pathogenic mutations in the hydrophobic region had any effect on the relative amounts of the transmembrane forms of PrP; known pathogenic mutations in other regions did not have any effect. This implies that transmembrane PrP might not play as big role as previously thought in the pathogenesis of prion diseases (Stewart and Harris 2001).

5 Using topology inversion for function

5.1 SecG

The bacterial protein conducting channel, SecYEG, consists of three inner membrane proteins that are responsible for the insertion of proteins into the membrane and the transportation of secretory proteins through the membrane.

There is some experimental evidence to suggest not only that SecG can undergo topology inversion (Nishiyama et al. 1996; Nagamori et al. 2002), but also that this is not necessary for the function of the SecYEG complex (van der Sluis et al. 2006). Modified SecG molecules that have a single topology, either by cysteine cross-linking or by gene-fusions, are indeed functional, but the rate of protein translocation is higher if SecG is unrestricted to alter its topology. However, for this to occur, the soluble protein SecA is required and there is also suggestion that other factors such as SecDF may be required to invert SecG (Sugai et al. 2007).

6 Using dual-topology as a targeting system

There are indications that some membrane proteins exhibiting dual-topology in the ER use their different forms as a way of targeting themselves to the correct membrane in some tissues. Interestingly, both proteins that are reported to do this are expressed in hepatocyte cells in the liver.

6.1 Cytochrome p450-2E1

Cytochrome p450s (p450s) has been implicated in alcoholic liver disease and is most commonly found in the ER where it is involved in the metabolism of various small hydrophobic compounds. It consists of one N-terminal hydrophobic transmembrane region and a large catalytic C-terminal domain. This domain is normally exposed to the cytoplasmic side of the ER membrane. However, immunofluorescent microscopy has indicated that some p450s have their catalytic domain on the outside of hepatocyte cells in rat and human livers (Wu and Cederbaum 1992). Neve and Ingelman-Sundberg (2000) showed that for one specific p450s, CYP2E1, the C-terminal domain is located on the outside of the plasma membrane and that transport from the ER to the plasma membrane is dependent on a small amount (~2%) of CYP2E1 being inserted with opposite topology in the ER membrane during translation. These are then transported via the Golgi apparatus to the outer surface of the plasma membrane.

6.2 Epoxide hydrolase

Epoxide hydrolase is involved in the metabolism of xenobiotics and in the hepatocyte uptake of bile acid. It is expressed in the ER in two opposite orientations and most likely has three transmembrane helices. One of the topologies is then transported to the plasma membrane of COS-7 cells (Zhu et al. 1999). Alterations of the number of positive charges in the N-terminal region lead to dramatic differences in inversions of the topology indicating that epoxide hydrolase is topologically sensitive, much like EmrE.

References

Appel N, Herian U, Bartenschlager R (2005) Efficient rescue of hepatitis C virus RNA replication by trans-complementation with nonstructural protein 5A. J Virol 79: 896–909

Bruss V and Vieluf K (1995) Functions of the internal pre-S domain of the large surface protein in hepatitis B virus particle morphogenesis. J Virol 69: 6652–6657

Bruss V, Lu X, Thomssen R, Gerlich WH (1994) Post-translational alterations in transmembrane topology of the hepatitis B virus large envelope protein. EMBO J 13: 2273–2279

Chan CS, Zlomislic MR, Tieleman DP, Turner RJ (2007) The TatA subunit of *Escherichia coli* twin-arginine translocase has an N-in topology. Biochemistry 46: 7396–7404

Chen YJ, Pornillos O, Lieu S, Ma C, Chen AP, Chang G (2007) X-ray structure of EmrE supports dual topology model. Proc Nat Acad Sci USA 104: 18999–19004

Choi S, Jeon J, Yang S, Kim S (2008) Common occurrence of internal repeat symmetry in membrane proteins. Proteins Struct Funct Bioinform 71: 68–80

Chung YJ and Saier MH Jr (2001) SMR-type multidrug resistance pumps. Curr Opin Drug Discov Dev 4: 237–245

DeLano WL (2002) The PyMOL molecular graphics system. DeLano Scientific. Palo Alto, USA

Dunlop J, Jones PC, Finbow ME (1995) Membrane insertion and assembly of ductin: a polytopic channel with dual orientations. EMBO J 14: 3609–3616

Finn RD, Tate J, Mistry J, Coggill PC, Sammut SJ, Hotz HR, Ceric G, Forslund K, Eddy SR, Sonnhammer EL, Bateman A (2008) The Pfam protein families database. Nucleic Acids Res 36: D281–D288

Gouffi K, Gerard F, Santini CL, Wu LF (2004) Dual topology of the *Escherichia coli* TatA protein. J Biol Chem 279: 11608–11615

Granseth E, von Heijne G, Elofsson A (2005) A study of the membrane–water interface region of membrane proteins. J Mol Biol 346: 377–385

Hegde RS, Mastrianni JA, Scott MR, DeFea KA, Tremblay P, Torchia M, DeArmond SJ, Prusiner SB, Lingappa VR (1998a) A transmembrane form of the prion protein in neurodegenerative disease. Science 279: 827–834

Hegde RS, Voigt S, Lingappa VR (1998b) Regulation of protein topology by trans-acting factors at the endoplasmic reticulum. Mol Cell 2: 85–91

Hegde RS, Tremblay P, Groth D, DeArmond SJ, Prusiner SB, Lingappa VR (1999) Transmissible and genetic prion diseases share a common pathway of neurodegeneration. Nature 402: 822–826

Kimura T, Ohnuma M, Sawai T, Yamaguchi A (1997) Membrane topology of the transposon 10-encoded metal-tetracycline/H+ antiporter as studied by site-directed chemical labeling. J Biol Chem 272: 580–585

Krogh A, Larsson B, von Heijne G, Sonnhammer EL (2001) Predicting transmembrane protein topology with a hidden Markov model: application to complete genomes. J Mol Biol 305: 567–580

Lambert C and Prange R (2001) Dual topology of the hepatitis B virus large envelope protein: determinants influencing post-translational pre-S translocation. J Biol Chem 276: 22265–22272

Lambert C, Mann S, Prange R (2004) Assessment of determinants affecting the dual topology of hepadnaviral large envelope proteins. J Gen Virol 85: 1221–1225

Lundin M, Lindstrom H, Gronwall C, Persson MA (2006) Dual topology of the processed hepatitis C virus protein NS4B is influenced by the NS5A protein. J Gen Virol 87: 3263–3272

Manoil C (1991) Analysis of membrane protein topology using alkaline phosphatase and beta-galactosidase gene fusions. Methods Cell Biol 34: 61–75

Nagamori S, Nishiyama K, Tokuda H (2002) Membrane topology inversion of SecG detected by labeling with a membrane-impermeable sulfhydryl reagent that causes a close association of SecG with SecA. J Biochem 132: 629–634

Neve EP, Ingelman-Sundberg M (2000) Molecular basis for the transport of cytochrome P450 2E1 to the plasma membrane. J Biol Chem 275: 17130–17135

Nilsson J, Persson B, von Heijne G (2005) Comparative analysis of amino acid distributions in integral membrane proteins from 107 genomes. Proteins 60: 606–616

Nishino K and Yamaguchi A (2001) Analysis of a complete library of putative drug transporter genes in *Escherichia coli*. J Bacteriol 183: 5803–5812

Nishiyama K, Suzuki T, Tokuda H (1996) Inversion of the membrane topology of SecG coupled with SecA-dependent preprotein translocation. Cell 85: 71–81

Oates J, Barrett CM, Barnett JP, Byrne KG, Bolhuis A, Robinson C (2005) The *Escherichia coli* twin-arginine translocation apparatus incorporates a distinct form of TatABC complex, spectrum of modular TatA complexes and minor TatAB complex. J Mol Biol 346: 295–305

149

Ostapchuk P, Hearing P, Ganem D (1994) A dramatic shift in the transmembrane topology of a viral envelope glycoprotein accompanies hepatitis B viral morphogenesis. EMBO J 13: 1048–1057

Parks GD, Hull JD, Lamb RA (1989) Transposition of domains between the M2 and HN viral membrane proteins results in polypeptides which can adopt more than one membrane orientation. J Cell Biol 109: 2023–2032

Petaja-Repo UE, Hogue M, Laperriere A, Bhalla S, Walker P, Bouvier M (2001) Newly synthesized human delta opioid receptors retained in the endoplasmic reticulum are retrotranslocated to the cytosol, deglycosylated, ubiquitinated, and degraded by the proteasome. J Biol Chem 276: 4416–4423

Prange R and Streeck RE (1995) Novel transmembrane topology of the hepatitis B virus envelope proteins. EMBO J 14: 247–256

Prusiner SB (1997) Prion diseases and the BSE crisis. Science 278: 245–251

Rapp M, Granseth E, Seppala S, von Heijne G (2006) Identification and evolution of dual-topology membrane proteins. Nat Struct Mol Biol 13: 112–116

Rapp M, Seppala S, Granseth E, von Heijne G (2007) Emulating membrane protein evolution by rational design. Science 315: 1282–1284

Schuldiner S, Granot D, Mordoch SS, Ninio S, Rotem D, Soskin M, Tate CG, Yerushalmi H (2001) Small is mighty: EmrE, a multidrug transporter as an experimental paradigm. News Physiol Sci 16: 130–134

Sebag JA and Hinkle PM (2007) Melanocortin-2 receptor accessory protein MRAP forms antiparallel homodimers. Proc Nat Acad Sci USA 104: 20244–20249

Sebag JA and Hinkle PM (2009) Regions of melanocortin 2 (MC2) receptor accessory protein necessary for dual topology and MC2 receptor trafficking and signaling. J Biol Chem 284: 610–618

Shimizu T, Mitsuke H, Noto K, Arai M (2004) Internal gene duplication in the evolution of prokaryotic transmembrane proteins. J Mol Biol 339: 1–15

Soskine M, Mark S, Tayer N, Mizrachi R, Schuldiner S (2006) On parallel and antiparallel topology of a homodimeric multidrug transporter. J Biol Chem 281: 36205–36212

Stewart RS and Harris DA (2001) Most pathogenic mutations do not alter the membrane topology of the prion protein. J Biol Chem 276: 2212–2220

Sugai R, Takemae K, Tokuda H, Nishiyama K (2007) Topology inversion of SecG is essential for cytosolic SecA-dependent stimulation of protein translocation. J Biol Chem 282: 29540–29548

Thompson JD, Higgins DG, Gibson TJ (1994) CLUSTAL W: improving the sensitivity of progressive multiple sequence alignment through sequence weighting, position-specific gap penalties and weight matrix choice. Nucleic Acids Res 22: 4673–4680

van der Sluis EO, van der Vries E, Berrelkamp G, Nouwen N, Driessen AJ (2006) Topologically fixed SecG is fully functional. J Bacteriol 188: 1188–1190

von Heijne G (2006) Membrane-protein topology. Nat Rev Mol Cell Biol 7: 909–918

Wallin E and von Heijne G (1998) Genome-wide analysis of integral membrane proteins from eubacterial, archaean, and eukaryotic organisms. Protein Sci 7: 1029–1038

Weiner JH, Bilous PT, Shaw GM, Lubitz SP, Frost L, Thomas GH, Cole JA, Turner RJ (1998) A novel and ubiquitous system for membrane targeting and secretion of cofactor-containing proteins. Cell 93: 93–101

Wu D and Cederbaum AI (1992) Presence of functionally active cytochrome P-450IIE1 in the plasma membrane of rat hepatocytes. Hepatology 15: 515–524

Zhu Q, von Dippe P, Xing W, Levy D (1999) Membrane topology and cell surface targeting of microsomal epoxide hydrolase. Evidence for multiple topological orientations. J Biol Chem 274: 27898–27904

Predicting the burial/exposure status of transmembrane residues in helical membrane proteins

Volkhard Helms, Sikander Hayat and Jennifer Metzger

Center for Bioinformatics, Saarland University, Saarbruecken, Germany

Abstract

In multipass transmembrane proteins one face of the transmembrane helices is in contact with the aliphatic acyl chains of the phospholipids and with the polar interface region. The other face makes contacts with other helices or points into the protein interior. In larger proteins, some helices may even be buried completely. Analysis of the available three-dimensional crystal structures has shown that inwards pointing residues tend to be more conserved than outwards pointing residues. Furthermore, residues pointing outwards are generally very hydrophobic whereas inward pointing residues may have different characteristics. Based on these two findings, knowledge-based propensity scales have been derived that, when combined with analysis of residue conservation, allow predicting the exposure status of residues in the hydrophobic core region with about 80% accuracy. These tools give biologists insight in the putative topology of transmembrane helix bundles.

1 Introduction

Bitopic membrane proteins contain a single alpha-helix that crosses the lipid bilayer once. In polytopic transmembrane (TM) proteins, the amino acid chain traverses the lipid bilayer multiple times. To satisfy the hydrogen-bonding requirements of the polar backbone atoms, polytopic membrane proteins either adopt the topology of alpha-helical bundles in their membrane part or that of beta-barrels. In this chapter, we will focus on the topologies of helical membrane proteins. Ideally, these occur as helical bundles where a number of TM helices in perfect straight conformation cross

Corresponding author: Volkhard Helms, Center for Bioinformatics, Saarland University, 66041 Saarbruecken, Germany (E-mail: volkhard.helms@bioinformatik.uni-saarland.de)

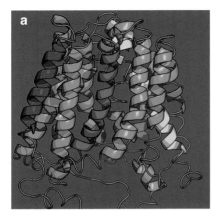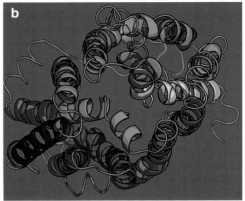

Fig. 1. (a) Shows a side view and (b) a top view of the 3D structure representation of lactose permease (pdb id: 2cfq). TM helix 1 is colored blue, TM helix 10 is colored yellow, and the last TM helix 12 is colored red. TM helices 1 and 12 are fully exposed, TM helix 10 is partially buried in its lower half.

the lipid bilayer as a whole. Figure 1 shows the three-dimensional (3D) structure of the membrane transporter lactose permease, which will be used as an educational example in this chapter. Obviously, its TM helices are not perfectly straight. Still, this structure conforms quite well to the canonical picture of a helical bundle.

When we are given the sequence of a helical membrane protein, it is quite straight-forward to identify the membrane-spanning segments from the amino acid sequence with modern bioinformatics techniques (see the chapter by Casadio et al., this volume). Although the topologies of beta-barrels are even more regular than those of helical proteins, bioinformatics detection of beta strands from a protein sequence is more difficult, since the exterior of the beta-barrel proteins is less hydrophobic than that of alpha-helical TM proteins. Once the strands are known, assembling them into a barrel form is not that much of a problem anymore. The only remaining challenge is predicting the register shift between adjacent beta strands (Jackups and Liang 2005). In contrast, making structural predictions for helical bundles beyond the identification of TM helices is quite hard because the 3D structures in this class of proteins have turned out to be far less homogenous. Some methods that predict the re-entrant loop regions and helix kinks have been discussed in (the chapter by Granseth, this volume).

One of the simplest properties beyond the identification of helical segments is to predict whether a TM helix faces the lipids at all, or whether it is buried in the protein. Recent sequence-based bioinformatics approaches allow making predictions about the buried/exposure status of single residues in TM helices and these will be

In order to quantify the level of conservation one needs a MSA with at least 20 related protein sequences to accurately estimate conservation indices. Various methods exist for computing conservation indices. We have worked with the program Al2Co presented in (Pei and Grishin 2001). In the variance-based measure, the conservation index $C(i)$ for sequence position i in an MSA compares $f_j(i)$, namely the

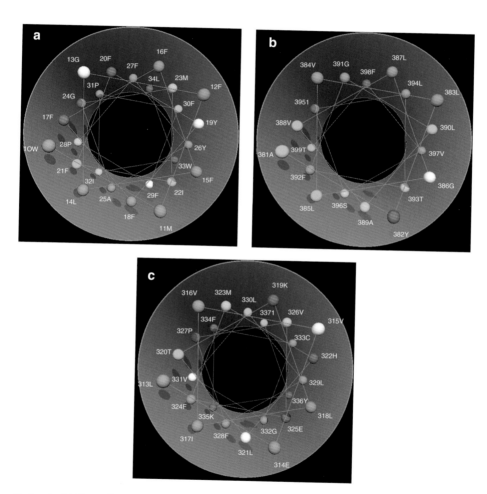

Fig. 3. (a, b) Show the conservation indices of the residues in helices 1 and 12 of lactose permease (pdb id: 2cfq). (c) Shows the conservation indices of residues in helix 10. In the crystal structure of the protein, residues exposed to the bilayer are less conserved than the residues buried in the protein structure. The color scale is slate blue for residues with conservation index (cl) <-0.3, Cyan for $-0.3 \leq cl < 0.0$, quartz for $0.0 \leq cl < 0.3$, neon pink for $0.3 \leq cl < 0.8$, and dark purple for $0.8 \leq cl$. The figures were generated with the tool TopoView (http://gepard.bioinformatik.uni-saarland.de).

frequency of amino acid j in the sequence position i to the overall frequency f_j of amino acid j in the alignment

$$C(i) = \sqrt{\sum_j (f_j(i) - f_j)^2}.$$

A position with $f_j(i)$ equal to f_j for all amino acids j is assigned $C(i) = 0$. On the contrary, $C(i)$ takes on its maximum for the position occupied by an invariant amino acid whose overall frequency in the alignment is low. Figure 3 shows the conservation indices of the positions along three TM helices of the lactose permease membrane transporter. Helices 1 and 12 are boundary helices whereas helix 10 is partly buried.

Beuming and Weinstein (2004) made a pioneering contribution when they combined amino acid propensities with sequence conservation. Following the approach of Beuming and Weinstein, we have combined in our TMX method (Park et al. 2007) the conservation index $C(i)$ and the frequency profile into a positional score $S(i)$ of each residue position i:

$$S(i) = C_C \cdot C(i) + \sum_{j=1}^{20} C_j \cdot f_j(i).$$

Here, C_C and C_j are coefficients that may be optimized by linear or nonlinear techniques. In the approach by Beuming and Weinstein, both coefficients were set to 0.5. The TMX method is a two-stage classifier that predicts each position i to be buried or exposed. In the first stage, the score for position i is computed using a window of size 21. This means that on both sides of position i, the frequency profiles of ten further residues are considered. The second stage then uses a binary support vector classifier to classify the positional score as exposed or buried. The advantage of this approach is that also a confidence value is provided for each prediction. The TMX method has an accuracy of 78.71% from a leave-one-out test. This is significantly higher than the 68.67% accuracy of the Beuming–Weinstein method when tested in the same manner on the same dataset and also higher than the 71.06% for the method of Teasdale et al. (Yuan et al. 2006) who developed a classifier to predict real-valued rSASA values. Figure 4 shows the TMX predictions for the same three helices of lactose permease. The clear separation of the red and blue faces of the helices in this graphical representation indicates that helices 1 and 12 are likely boundary helices whereas helix 10 is partly buried and partly exposed.

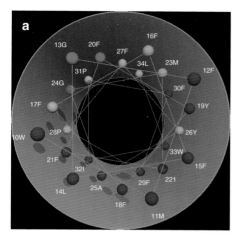
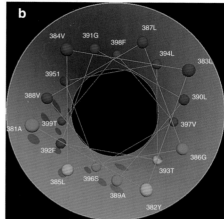
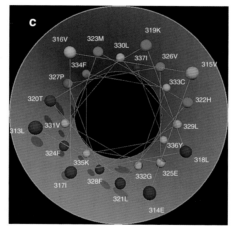

Fig. 4. **(a, b) Show the predicted burial status according to the TMX method for the residues in helices 1 and 12 of Lactose permease (pdb id: 2cfq). (c) Shows the burial status prediction results for helix 10.** Helices 1 and 12 show a uniform distinction between their buried and exposed faces, while for helix 10 the direction of residues exposed to the bilayer gradually changes for residues near the cytoplasmic side of the membrane to the residues near the periplasmic side. For residues predicted as buried by TMX, the color scale is orange for a predicted positional score $PS \leq 0.0$, orange red for $0.0 < PS \leq 0.5$ and red for $0.5 \leq PS$. So red residues are predicted as buried with highest confidence. For residues predicted as exposed, the color scale is cyan for $PS \leq 0.0$, slate for $0.0 < PS \leq 0.5$, and blue for $0.5 \leq PS$. There, blue residues are predicted as exposed with highest confidence.

The recent method RHYTHM by Hildebrand et al. is based on a new set of propensity scales for residues to contact another helix or the membrane (Rose et al. 2009). Separate scales were derived for a dataset of 21 channel proteins and for a

161

dataset of 14 "membrane-coils" proteins. Evolutionary information about sequence conservation is additionally derived from the MSAs stored in the Pfam database. Conservation is added as a bonus to the contact predictions. The authors noted that this Pfam prediction did not significantly affect the specificity of the predictions. So far, RHYTHM has not been benchmarked against the earlier methods.

5 Applications of burial prediction

There now exists several web-services that allow predicting buried versus exposed regions of TM helices, see Table 2. Whereas ProperTM, LIPS, RANTS, and TMX depend on generating MSAs to produce predictions about TM helix orientations or solvent accessibility, the method RHYTHM avoids this time-consuming step because it utilizes information from precomputed alignments.

An ideal area for applying bioinformatics techniques for burial/exposure prediction is the positioning and orientation of TM helices in density maps from electron microscopy (Fleishman et al. 2004a). Structure determination through cryoelectron microscopy typically results in intermediate- or low-resolution structural information. In these maps one can often clearly identify the positions of TM helices but not those of individual amino acids. Further experimental data or predictions of further structural constraints may help establishing the correspondence between the TM segments of the sequence and the densities in the EM map assumed to be TM helices. If this can be successfully done, computational tools for predicting the rotational angles of TM helices about the helix axes come into play. For example, in a study on the gap junction channel connexin, cryo-EM could resolve the positions and tilt angles of the four TM helices at 5.7 Å in-plane resolution and 19.8 Å vertical resolution (Fleishman et al. 2004b). There are obviously $4 \times 3 \times 2 \times 1 = 24$ possibilities to assign the four TM segments to the four helix positions. The authors have considered all those solutions where the evolutionarily variable positions of the helices are placed in lumen- or lipid-exposed positions, whereas conserved faces

Table 2. Available web-servers to predict the orientation of transmembrane helices

Name	URL	References
ProperTM	*http://icb.med.cornell.edu/services/propertm/start*	Beuming and Weinstein (2004)
LIPS	*http://gila.bioengr.uic.edu/lab/larisa/lips.html*	Adamian and Liang (2006b)
RANTS	*http://gila.bioengr.uic.edu/lab/larisa/rants.html*	Adamian and Liang (2006a)
TMX	*http://service.bioinformatik.uni-saarland.de/tmx/*	Park et al. (2007)
RHYTHM	*http://proteinformatics.charite.de/rhythm/*	Rose et al. (2009)

were packed inside the protein core. Moreover they could identify in the MSA of the connexin protein family five pairs of correlated mutations in the TM and juxtamembrane domains. Based on the underlying assumption that contacting pairs of residues undergo dependent evolution these residue pairs could be used as additional structural constraints where these residue pairs all need to make direct contacts to each other. These constraints then allowed the authors to optimize the rotational helical angles and suggest a unique structural model for the gap junction channel at residue resolution.

Another possible application of exposure/burial predicting techniques is the prediction of contact interfaces on the TM parts of the protein surfaces. In the area of soluble proteins, there is good evidence that residue conservation at protein interfaces is slightly higher than elsewhere on the protein surface. Also TM proteins often form oligomers or supracomplexes. Examples are the oligomerization of G-protein coupled receptors (Gurevich and Gurevich 2008) or the formation of supracomplexes in the respiratory chain (Wittig and Schägger 2009). Unfortunately, the number of known 3D structures and, in particular, that of oligomeric assemblies is still too low at the present time for a statistically meaningful analysis of this hypothesis.

As a last application, the prediction of transporter pores or of catalytic residues can be mentioned. In preliminary work we have shown that TMX profiles of transporter helices provide sufficient information to map the helices of transporters with unknown 3D structure in a threading-approach onto the pore-lining helices of transporters with known structure (Jan Christoph and Volkhard Helms, unpublished work).

In summary, we conclude that the prediction of buried/exposed states of TM residues in helical proteins has reached a mature state. Several robust methods have been presented that exploit the potential of propensity scales and of sequence conservation. New tools have recently started to additionally include structural information about helix contacts. There still seems to be room for improving the prediction accuracies by combining these novel structural features with the traditional sequence-based features in an optimal fashion. Also, one may consider developing hierarchical approaches that combine tools optimized for the subfamilies of membrane transporters, channels, receptors, etc.

References

Adamian L and Liang J (2006a) Prediction of buried helices in multispan alpha helical membrane proteins. Proteins 63: 1–5

Adamian L and Liang J (2006b) Prediction of transmembrane helix orientation in polytopic membrane proteins. BMC Struct Biol 6: 13

Adamian L, Nanda V, DeGrado WF, Liang J (2005) Empirical lipid propensities of amino acid residues in multispan alpha helical membrane proteins. Proteins 59: 496–509

163

Beuming T and Weinstein H (2004) A knowledge-based scale for the analysis and prediction of buried and exposed faces of transmembrane domain proteins. Bioinformatics 20: 1822–1835

Cronet P, Sander C, Vriend G (1993) Modeling the transmembrane seven helix bundle. Protein Eng 6: 59–64

Eisenberg D, Weiss RM, Terwilliger TC (1982) The helical hydrophobic moment: a measure of the amphiphilicity of a helix. Nature 299: 371–374

Eisenberg D, Schwarz E, Komaromy M, Wall R (1984) Analysis of membrane and surface protein sequences with the hydrophobic moment plot. J Mol Biol 179: 125–142

Engelman DM and Zaccai G (1980) Bacteriorhodopsin is an inside-out protein. Proc Nat Acad Sci USA 77: 5894–5898

Fleishman SJ, Harrington S, Friesner RA, Honig B, Ben-Tal N (2004a) An automatic method for predicting transmembrane protein structures using cryo-EM and evolutionary data. Biophys J 87: 3448–3459

Fleishman SJ, Unger VM, Yeager M, Ben-Tal N (2004b) A C-alpha model for the transmembrane alpha helices of gap junction intercellular channels. Mol Cell 15: 879–888

Gurevich VV and Gurevich EV (2008) GPCR monomers and oligomers: it takes all kinds. Trends Neurosci 31: 74–81

Jackups R Jr and Liang J (2005) Interstrand pairing patterns in b-barrel membrane proteins: the positive-outside rule, aromatic rescue, and strand registration prediction. J Mol Biol 354: 979–993

Park Y and Helms V (2006) How strongly do sequence conservation patterns and empirical scales correlate with exposure patterns of transmembrane helices of membrane proteins? Biopolymers 83: 389–399

Park Y and Helms V (2007) On the derivation of propensity scales for predicting exposed transmembrane residues of helical membrane proteins. Bioinformatics 23: 701–708

Park Y, Hayat S, Helms V (2007) Prediction of the burial status of transmembrane residues of helical membrane proteins. BMC Bioinform 8: 302

Pei J and Grishin NV (2001) AL2CO: calculation of positional conservation in a protein sequence alignment. Bioinformatics 17: 700–712

Pilpel Y, Ben-Tal N, Lancet D (1999) kPROT: a knowledge-based scale for the propensity of residue orientation in transmembrane segments. Application to membrane protein structure prediction. J Mol Biol 294: 921–935

Rees DC, DeAntonio L, Eisenberg D (1989) Hydrophobic organization of membrane proteins. Science 245: 510–513

Rose A, Lorenzen S, Goede A, Gruening B, Hildebrand PW (2009) RHYTHM – server to predict the orientation of transmembrane helices in channels and membrane-coils. Nucleic Acids Res 37: W575–W580

Stevens TJ and Arkin IT (1999) Are membrane proteins "inside-out" proteins? Proteins 36: 135–143

Taylor WR, Jones DT, Green NM (1994) A method for alpha-helical integral membrane protein fold prediction. Proteins 18: 281–294

Wittig I and Schägger H (2009) Supramolecular organization of ATP synthase and respiratory chain in mitochondrial membranes. Biochim Biophys Acta – Bioenerg 1787: 672–680

Yuan Z, Zhang F, David MJ, Boden M, Teasdale RD (2006) Predicting the solvent accessibility of transmembrane residues from protein sequence. J Proteome Res 5: 1063–1070

Helix–helix interaction patterns in membrane proteins

Dieter Langosch[1], Jana R. Herrmann[1], Stephanie Unterreitmeier[1] and Angelika Fuchs[2]

[1]Department für biowissenschaftliche Grundlagen, Technische Universität München and Center for Integrated Protein Science (CIPS^M), Freising, Germany
[2]Department für biowissenschaftliche Grundlagen, Technische Universität München, Freising, Germany

Abstract

Membrane-spanning α-helices represent major sites of protein–protein interaction in membrane protein oligomerization and folding. As such, these interactions may be of exquisite specificity. Specificity often rests on a complex interplay of different types of residues forming the helix–helix interfaces via dense packing and different non-covalent forces, including van der Waal's forces, hydrogen bonding, charge–charge interactions, and aromatic interactions. These interfaces often contain complex residue motifs where the contribution of constituent amino acids depends on the context of the surrounding sequence. Moreover, transmembrane helix–helix interactions are increasingly recognized as being dynamic and dependent on the functional state of a given protein.

Abbreviations: GpA, glycophorin A; H-bond, hydrogen bond; TMD, transmembrane domain.

1 Introduction

Studying membrane protein structure and assembly has made it clear that interactions and dynamics of α-helical transmembrane domains (TMDs) play a crucial role in their folding, oligomeric assembly, and function. Various aspects around this topic have been covered by excellent recent reviews (Fleming 2000; Popot and Engelman 2000; Shai 2001; Ubarretxena-Belandia and Engelman 2001; Arkin 2002; Helms 2002; Langosch et al. 2002; Chamberlain et al. 2003; DeGrado et al. 2003; Schneider

Corresponding author: Dieter Langosch, Department für biowissenschaftliche Grundlagen, Technische Universität München, Weihenstephaner Berg 3, and Munich Center For Integrated Protein Science (CIPS^M), 85354 Freising, Germany (E-mail: langosch@tum.de)

2004; Seelig 2004; Bowie 2005; MacKenzie 2006; Matthews et al. 2006; Rath et al. 2007, 2009; MacKenzie and Fleming 2008; Moore et al. 2008; Slivka et al. 2008; Langosch and Arkin 2009).

The importance of transmembrane helix–helix interactions for membrane protein folding was originally indicated by showing that the polytopic light-sensor bacteriorhodopsin could be split proteolytically into several fragments, which could subsequently be reassembled to functional protein (Popot et al. 1986; Ozawa et al. 1997). A role for TMD–TMD interactions in the non-covalent assembly of single-spanning, or bitopic, membrane proteins was demonstrated when the TMD of the major erythrocyte membrane protein glycophorin A (GpA) formed dimers on SDS gels with exquisite sequence-specificity (Bormann et al. 1989; Lemmon et al. 1992a,b). These findings were conceptualized in the two-stage model. In the first stage, transmembrane α-helices are membrane-integrated independent from each others and assemble via sequence-specific helix–helix interactions in the second stage (Popot and Engelman 1990, 2000).

TMD–TMD assembly results in distinct patterns of residue conservation during evolution. Specifically, TMDs of bitopic proteins are more conserved than the remainder of the protein and conservation is stronger at one side of the helix (Zviling et al. 2007). With polytopic proteins, sequence variation is higher where TMD helices face the lipid bilayer than at helix–helix interfaces (Samatey et al. 1995; Stevens and Arkin 2001). Further, single-spanning membrane proteins are more tolerant to mutation in comparison to multi-spanning proteins, where most TMDs contact multiple helices (Jones et al. 1994a,b). Together, this reflects conservation of amino acids at the sites of TMD–TMD packing and highlights their importance for specific interaction. Analyzing high-resolution structures of polytopic proteins showed preferential orientation of aliphatic residue types (Ile, Leu, Phe, and Val) toward the lipid phase while polar residues tend to participate in helix–helix interfaces (Liang et al. 2005). Small and hydroxylated residues (Gly, Ala, Ser, and Thr) prefer regions of high packing density (Adamian and Liang 2001). Neighboring pairs of residues with a high propensity of occurrence include Gly pairs, pairs of an aromatic residue and a basic residue (e.g., Trp–Arg, Trp–His, and Tyr–Lys), of polar non-ionizable residues (e.g., Asn–Asn, Gln–Asn, and Ser–Ser), of two ionizable residues, and of one ionizable residue and a residue with a carboxamide side chain (e.g., Asp–Asn, Javadpour et al. 1999; Adamian and Liang 2001). These contact potentials clearly point at a rich diversity of molecular forces within transmembrane helix–helix interfaces discussed in detail below. They also hint at the mechanisms that provide sequence-specificity of interaction. Nevertheless, we currently only have a rudimentary understanding of the mechanisms that ensure specificity of TMD–TMD interactions and avoidance of promiscuous ones. In addition, it is clear that these

166

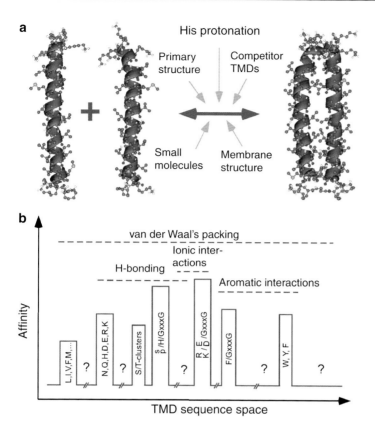

Fig. 1. **Molecular basis of transmembrane helix–helix assembly.** (**a**) Overview of factors that are known to influence TMD–TMD interaction. (**b**) A simplified depiction of how different types of interfacial residues and motifs might be distributed in TMD sequence space. The baseline of the distribution corresponds to low-affinity non-specific interactions; peak heights are crude estimates based on published data of model cases. Single letter designation for amino acids is used, s and p refer to small and polar, respectively. Part a is modified after Fig. 1 in Langosch and Arkin (2009).

interactions are frequently regulatable by expression of competitor sequences, side-chain protonation, lipid bilayer structure, small molecules, etc. (Fig. 1a). Regulating reversible interactions within the membrane is likely to be essential for regulation of protein function. Also, certain TMDs exhibit more than one interface in a complex, rendering it janus-headed (Rath et al. 2006; Barwe et al. 2007).

2 Technical approaches to identify transmembrane helix–helix interfaces

High-resolution membrane protein structures, and by implication of TMD–TMD interfaces, are experimentally investigated mostly by X-ray crystallography which has revealed the structures of about 210 unique polytopic membrane proteins (*http://blanco.biomol.uci.edu/Membrane_Proteins_xtal.html*) that currently account for only ~2% of all protein structures. Progress has been slower with bitopic proteins. While the structure of the GpA TMD dimer has already been solved over 10 years ago by NMR studies in detergent (MacKenzie et al. 1997) and later in membranes (Smith et al. 2001), about half a dozen NMR structures have been presented more recently and X-ray crystallography has solved one of them (Oxenoid and Chou 2005; Call et al. 2006; Bocharov et al. 2007, 2008a,b; Schnell and Chou 2008; Stouffer et al. 2008; Lau et al. 2009; Sato et al. 2009; Stein et al. 2009; Wang et al. 2009; Yang et al. 2009).

High-resolution structures provide detailed insight into protein–protein interfaces but do not necessarily identify the most critical residues that may form "hotspots" of interaction within them. Patterns of interfacial amino acids have also been identified by biochemical and biophysical methods that measure non-covalent TMD–TMD assembly coupled to point mutagenesis. Assembly may be examined by gel shift assays, analytical ultracentrifugation, fluorescence resonance transfer, and disulfide exchange in detergent or membranes (reviewed in: Ridder and Langosch 2005; MacKenzie 2006; Merzlyakov et al. 2007; Fleming 2008; Merzlyakov and Hristova 2008). In addition, genetic approaches have been developed where interaction is monitored in a natural membrane environment. These genetic approaches allow investigation of candidate TMDs. In addition, they also permit exploration of TMD–TMD interfaces in a systematic ab initio approach by selection of self-interacting TMDs from combinatorial libraries of randomized hydrophobic sequences (Russ and Engelman 2000; Gurezka and Langosch 2001; Dawson et al. 2002; Ridder and Langosch 2005; Unterreitmeier et al. 2007; Herrmann et al. 2009, 2010). Selection of high-affinity TMDs requires an experimental system where their interaction results in a selectable phenotype. The ToxR transcription activator system has been developed for this purpose (Langosch et al. 1996) and exploits the fact that self-interaction of ToxR-embedded TMDs within the inner membrane of expressing *Escherichia coli* reporter strains enhances expression of chloramphenicol resistance. The ToxR system exists in two versions used for library screening for homotypic interactions, TOXCAT (Russ and Engelman 1999) and POSSYCCAT (Gurezka and Langosch 2001). It has been modified to investigate heterotypic interactions in a dominant-negative fashion (Lindner and Langosch 2006; Yin et al. 2007; Herrmann et al. 2009). The beauty of the library screening approach is that interfacial consensus motifs emerge from alignments of selected sequences and can be verified by mutational analysis and reconstruction on neutral host sequences. Moreover, searching homology-purged databases can reveal whether or not a given motif is overrepresented in natural TMDs. Overrepresented motifs are likely to infer a

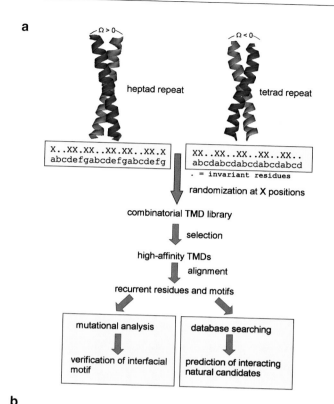

a

```
X..XX.XX..XX.XX..XX.X
abcdefgabcdefgabcdefg
```

```
XX..XX..XX..XX..XX..
abcdabcdabcdabcdabcd
. = invariant residues
```

heptad repeat

tetrad repeat

randomization at X positions

combinatorial TMD library

selection

high-affinity TMDs

alignment

recurrent residues and motifs

mutational analysis

verification of interfacial motif

database searching

prediction of interacting natural candidates

b

interfacial motif:	variant aa	invariant aa	reference
pa..Ga..Ga..T	G, A, V, L, I, S, T, P, A	L	1
.ᵖ..aG..aG..a	G, A, V, L, I, S, T, P, A	A	1
....F..G...G....	A, C, F, G, I, L, M, R, S, T, V, W	A	2
.ᵖ..ᵖH......G...G	all natural residues	L	3
....DR......G..G	all natural residues	L	4
.....R.E....G..G	all natural residues	L	4
....S..SS..T	A, T, S, F, V, L, I, P	L	5
...S...SS..T	A, T, S, F, V, L, I, P	L	5
W......W......W.	all natural residues	A	6

p = polar (S, T)
a = aliphatic (L, I, V)
s = small (G, A, S)
. = residues outside consensus motif

Fig. 2. **Approaches and outcomes in screening combinatorial libraries for high-affinity TMDs.** (**a**) Outline of library construction and screening. The outcome of individual screens depends on whether tetrad or heptad motifs are randomized, on the hydrophobicity of invariant amino acids, and on the complement of codons used for the variant ones. (**b**) Recurrent motifs as identified from different libraries where different interfacial residue patterns had been randomized with different sets of amino acids on different invariant host backgrounds. Ω = helix/helix crossing angle; aa = amino acid. The presence of GxxxG motifs in high-affinity TMDs suggests that the corresponding helix–helix pairs have negative crossing angles, even though a heptad-repeat pattern underlying left-handed pairs had been randomized. References: 1 – Russ and Engelman (2000); 2 – Unterreitmeier et al. (2007); 3 – Herrmann et al. (2009); 4 – Herrmann et al. (2010); 5 – Dawson et al. (2002); 6 – Ridder et al. 2005). Modified after Langosch and Arkin (2009), Fig. 2.

functional advantage for the proteins in question, for example via stronger oligomer formation. Also, database searching leads to testable predictions of related motifs in natural membrane proteins. Figure 2a illustrates the general strategy, while Fig. 2b summarizes the results obtained so far.

3 Structure of transmembrane helix–helix interfaces

The structure of TMD–TMD interfaces is both defined by the geometry of side-chain packing and by more focal forces, like hydrogen bonding (H-bonding), charge–charge interactions, and aromatic interactions. These different forces frequently cooperate to form complex interfaces that exhibit high degrees of sequence-specificity. As a result, the role of individual amino acids tends to be highly dependent on the context of the surrounding structure. In the following, those different forces are discussed separately for the sake of simplicity. Figure 1b summarizes how different interfacial amino acid motifs may be distributed in sequence space.

3.1 Amino acid side-chain packing

Due to packing constraints, the long axes of soluble or transmembrane helix–helix pairs usually adopt either positive or negative crossing angles (Chothia 1984; Bowie 1997). In a simplified model, interfacial residues of pairs with positive crossing angles, also termed left-handed pairs, follow a $[a..de.g]_n$ heptad-repeat pattern (where lower case letters represent residue positions) reminiscent of leucine zippers where side chains of one helix form "knobs" that pack into "holes" of the opposite helix surface. The interfaces of pairs characterized by negative angles, or right-handed pairs, correspond to a $[ab..]_n$ tetrad repeat where side-chain packing is less regular than in the "knobs-into-holes" model (Langosch and Heringa 1998; Langosch et al. 2002, Fig. 2a). Accordingly, a recent rigorous structural classification of TMD–TMD pairs from polytopic proteins revealed that about 2/3 of them fall into only four structural clusters, i.e., antiparallel and parallel helices with a limited range of crossing angles that is dictated by the nature of side-chain interactions (Walters and deGrado 2006). This suggests a limited conformation space for TMD–TMD pairs, as predicted based on geometrical considerations (Oberai et al. 2006). However, it has to be borne in mind that the remaining third of these pairs correspond to additional conformations with more varied crossing angles and irregularities in helix structures (mostly wide or tight helical turns that are often associated with kinks, Lehnert et al. 2004). The same broad structural classification seems to hold true for TMD–TMD assemblies from bitopic proteins as indicated by high-resolution structures (MacKenzie et al. 1997; Smith et al. 2001; Call et al. 2006; Bocharov et al. 2007, 2008b; Schnell and Chou 2008; Stouffer et al. 2008) and scanning mutagenesis

(Laage and Langosch 1997; Li et al. 2004b; Ruan et al. 2004a,b; Sulistijo and MacKenzie 2006; Dews and MacKenzie 2007).

The formation of well-packed interfaces is supported by non-directional van der Waal's forces. Albeit weak, undirectional, and strongly dependent on distance, van der Waal's interactions apply to any type of side-chain atom and accumulate over an entire well-packed interface. As such, they are suited to support interaction of TMDs composed of aliphatic residues, such as oligo-Leu helices (Gurezka et al. 1999; Mall et al. 2001; Ash et al. 2004). Similarly, TMDs containing only Leu, Ile, Val, Met, and Phe arranged in a heptad-repeat pattern tend to self-interact with little sequence-specificity, yet are overrepresented in natural bitopic membrane proteins (Gurezka and Langosch 2001). The NMR structure of the ErbB2 TMD (pdb code: 2jwa) dimer provides an example where an interface is primarily composed of non-polar side chains plus a Gly residue at the site of closest contact. The observation that fluorinated interfaces enhance interaction of TMD helices (Naarmann et al. 2006) could be explained by polarization of neighboring side-chain atoms by fluorine.

3.2 GxxxG motifs

GxxxG motifs exist in many TMDs and can induce their interaction. As their function appears to result from different physical forces, they are discussed in this separate chapter. A GxxxG motif has first been seen when interfacial residues of the GpA TMD–TMD homodimer (Lemmon et al. 1992a,b, 1994; Langosch et al. 1996; Fleming et al. 1997; Fisher et al. 1999; Russ and Engelman 1999; Fleming and Engelman 2001; Doura and Fleming 2004; Doura et al. 2004) were mapped by mutagenesis. Identification of the GxxxG motif as such was originally based on the observation that changing the residue spacing between both Gly residues affects dimerization and GxxxG induces self-interaction of model TMDs (Brosig and Langosch 1998). The GpA TMD dimer exhibits a negative crossing angle as implied by molecular modeling (Treutlein et al. 1992; Adams et al. 1996) and confirmed by NMR studies (MacKenzie et al. 1997; Smith et al. 2001). The contribution of GxxxG to an interface is apparently driven by a complex mixture of attractive forces and entropic factors (MacKenzie and Engelman 1998). It has been suggested that it leads to formation of a flat helix surface that maximizes van der Waal's interactions and that the loss of side-chain entropy upon association is minimal for Gly (Russ and Engelman 2000). Moreover, the Gly residues reduce the distance between the helix axes and thus may facilitate hydrogen bond formation between their C_α-hydrogens and the backbone carbonyl of the partner helix (Senes et al. 2001a). The early work on GpA TMD assembly was particularly rewarding since the GxxxG motif and degenerate versions thereof (designated "smallxxxsmall" or "GxxxG-like" with Gly exchanged for Ala, Ser, Cys, etc.), were later found in many other TMDs, including those of syndecans (Asundi and Carey 1995; Dews and MacKenzie 2007), members of the BNIP family (Sulistijo and MacKenzie 2006, 2009; Bocharov et al. 2007), protein tyrosine phosphatases (Chin et al. 2005), viral envelope proteins (Miyauchi et al. 2005; Arbely et al. 2006), growth factor receptors (Mendrola et al. 2002; Bocharov et al. 2008b), integrins (Gottschalk et al. 2002; Schneider and Engelman 2004; Lin et al. 2006a; Slivka et al. 2008; Wegener and Campbell 2008), and the Alzheimer precursor protein (Kim et al. 2005; Munter et al. 2007; Gorman et al. 2008) where it occurs in tandem.

Screening combinatorial TMD libraries where a tetrad repeat pattern had been randomized yielded high-affinity GxxxG motifs in more than 80% of all isolates (Russ and Engelman 2000), thus underpinning the role of this motif in TMD–TMD interactions. Indeed, database searching identified the GxxxG motif as the most prevalent pair-wise motif in TMDs (Arkin and Brünger 1998; Senes et al. 2000; Unterreitmeier et al. 2007). Overrepresentation of GxxxG relative to statistical expectation demonstrates that its presence supports protein function in evolution. At the same time, the fact that 12.5% of TMDs from non-homologous bitopic proteins contain at least one GxxxG motif (Senes et al. 2000; Unterreitmeier et al. 2007) suggests that mechanisms must have evolved to prevent promiscuous interaction of TMDs with GxxxG. Indeed, the mere presence of such motifs does not reliably predict high-affinity interaction. This is exemplified by the fact that GxxxG present within the ErbB2 receptor TMD lies outside the interface, which extends only over the N-terminal half of the helix (Bocharov et al. 2008b). Indeed, the N-terminal half self-associates with slightly higher propensity than the C-terminal half. It was suggested that interaction of the former one stabilizes the active state of the receptor while the latter one forms an interface in the inactive state (Escher et al. 2009). Further, GxxxG is highly effective within the contexts of oligo-Met and oligo-Val sequences (Brosig and Langosch 1998), but not within either an oligo-Leu TMD, a number of randomized TMDs (Unterreitmeier et al. 2007) or the M13 major coat protein TMD (Johnson et al. 2006). To avoid promiscuous homo- and heterotypic interactions, the impact of GxxxG depends on sequence context (Melnyk et al. 2004). This is underpinned by the finding that the interaction energy of the GpA TMD varies over a wide range after mutation of the sequence surrounding GxxxG (Doura et al. 2004). High-affinity TMDs holding GxxxG may therefore be regarded as islands in GxxxG sequence space. Screening combinatorial TMD libraries has identified some of these islands by showing that GxxxG can form high-affinity interfaces with appropriately spaced Phe (Unterreitmeier et al. 2007), clusters of His and polar/small residues (Herrmann et al. 2009), or ionizable residues (Herrmann et al. 2010) as described below.

3.3 Hydrogen bonding

The role of H-bonds in TMD–TMD interfaces is discussed controversially. On one hand, polar residues were inferred to form extensive H-bond connections that enhance packing between the TMDs of polytopic membrane proteins (Adamian and Liang 2002). Also, Asn and Gln residues strongly promote self-interaction of model (Choma et al. 2000; Zhou et al. 2000, 2001; Gratkowski et al. 2001; Ruan et al. 2004b) or natural (Ruan et al. 2004a) TMDs. These polar residues are thought to form strong interhelical H-bonds within the apolar millieu of lipid bilayers. Apart from hydroxylated side chains and carboxamides, homotypic interaction of model TMDs is also promoted by ionizable residues, including Asp, Glu, His (Gratkowski et al. 2001; Zhou et al. 2001; Sal-Man et al. 2004), Lys, and Arg (Johnson et al. 2007), which may also be attributed to H-bond formation in the absence of an oppositely charged residue on the partner helix.

On the other hand, recent studies suggest only modest stabilization of a bitopic model TMD–TMD interface by H-bonds (North et al. 2006) and H-bonds seem to contribute little toward stability of bacteriorhodopsin in SDS micelles (Joh et al. 2008; see Grigoryan and Degrado 2008 for a discussion of these results). Apart from H-bonds contributed by polar side-chains, it has been proposed that the C_α–H group is capable of participating in H-bonding (Senes et al. 2001b) since the marginal polarity of the C_α proton might be sufficient to serve as an H-bond donor in a highly hydrophobic environment. However, the effect upon stability of a single C_α–H $\cdots$ O=C bond in bacteriorhodopsin was estimated by mutagenesis in detergent micelles to be insignificant (Yohannan et al. 2004) and the enthalpy of a similar H-bond in GpA is relatively small (0.88 kcal/mol; Arbely and Arkin 2004) compared to an H-bond extending from a polar side chain (~2–3 kcal/mol).

Thus, the extent to which an H-bond contributes to the stability of a given interface may critically depend on its structural environment. One example underscoring this notion is the finding that high-affinity TMDs isolated from a combinatorial library were enriched for His residues which were frequently accompanied by Gly, Ser, and/or Thr residues at positions i-4 and i-1 relative to His (Herrmann et al. 2009). Mutational analyses confirmed the importance of these residues in homotypic interaction. Probing heterotypic interactions indicated that His residues interact in trans with hydroxylated residues suggesting that hydrogen bonds and possibly aromatic interactions stabilize the interface. Interestingly, the sequences with the highest affinities contained a C-terminal GxxxG motif which results in a [G/S/T] xx[G/S/T]HxxxxxxGxxxG consensus pattern. Reconstruction of minimal interaction motifs on an oligo-Leu sequence supported the idea that His is part of a H-bonded node that may be brought into register by a distant GxxxG (Herrmann et al. 2009). Isolated His residues support the assembly of model TMDs much less efficiently (Zhou et al. 2001; Herrmann et al. 2009). This exemplifies one case where precise geometric positioning, apparently accomplished here by GxxxG, may be required for optimal stabilization of H-bonds at a distant site. Database searching yielded only few candidate TMDs holding this motif. One of them corresponds to the previously well-investigated BNIP3 TMD. BNIP3 is a Bcl-2 family pro-apoptotic protein that initiates hypoxia-induced cell death. The BNIP3 TMD forms a homodimer characterized by the motif SHxxAxxxGxxxG (Sulistijo et al. 2003; Sulistijo and MacKenzie 2006) and its NMR structure confirmed these interfacial residues in the right-handed pair of helices (Bocharov et al. 2007; Sulistijo and Mackenzie 2009). The BNIP3 TMD–TMD interface thus corresponds to one variant of the consensus motif identified in a library screen. Apart from stabilizing interaction of bitopic subunits, His is also important in interfaces between the helices of polytopic proteins as residue triplets containing His and Ser or Thr are strongly overrepresented there (Adamian et al. 2003).

The context dependence of H-bonds in TMD–TMD interfaces is also supported by the formation of interfaces containing Ser/Thr-clusters (Dawson et al. 2002) and QxxS-motifs (Sal-Man et al. 2005). Self-interacting TMDs with predominant SxxSSxxT and SxxxSSxxT motifs were isolated from a combinatorial library and point mutagenesis showed the requirement of a cooperative network of interhelical H-bonds while single Ser or Thr residues did not promote interaction (Dawson et al. 2002). A QxxS motif was found essential for homodimerization of the bacterial Tar-1 protein and is significantly overrepresented in a bacterial TMD database suggesting its wide-spread role in homodimerization (Sal-Man et al. 2005).

3.4 Charge–charge interactions

Early evidence for charge–charge, or ionic, interactions between TMDs came from studies that probed the location of helices within the membrane. There, pairs of positively charged Lys and negatively charged Asp residues one helical turn apart placed a model helix deeper in the membrane than other spacings of the two residues (Chin and von Heijne 2000). On the other hand, heterotypic interaction of a pair of helices containing either Glu or Lys within an oligo-Leu host sequence did not exceed that of homotypic interaction in liposomal membranes (Shigematsu et al. 2002). The contribution of ionic interactions to oligomeric assembly was also tested for a few natural proteins. One well-investigated system corresponds to the T-cell receptor complex that is composed of single-span subunits. Three basic residues are found in the TMDs of the $\alpha\beta$ heterodimeric receptor while a pair of acidic residues is present in the TMDs of each of the three associated CD3$\gamma\epsilon$, CD3$\delta\epsilon$, and $\zeta\zeta$ signaling homodimers. Assembly of the complete oligomer rests on interaction of one basic residue of the central $\alpha\beta$ receptor with a pair of acidic residues within any of the signaling modules. Precise geometrical positioning of oppositely charged TMD residues is required for T-cell receptor complex assembly. The underlying TMD–TMD interaction is highly residue specific as Arg and Lys of the $\alpha\beta$ receptor heterodimer or Asp and Glu of the associated signaling modules cannot be exchanged for other residues of the same charge without loss of assembly competence (Call and Wucherpfennig 2007). The solvent NMR structure of the $\zeta\zeta$ homodimeric signaling module provides some clues as to the structural basis of specificity. In this isolated pair of subunits, helix–helix interaction is stabilized by a disulfide bond and the interface contains a H-bond between Tyr and Thr residues. In addition, one Asp side-chain oxygen of each helix forms an interhelical hydrogen bond to a carbonyl of the opposing strand while one seems to be available for interaction with a basic residue of the receptor. The presence of structural water within the $\zeta\zeta$ interface may precisely orient the ionizable side-chains within a network of H-bonds and thus explain residue specificity in charge–charge interaction (Call et al. 2006). A triad of basic and acidic residues also appears to drive assembly of a number of other activating immune receptors (Call and Wucherpfennig 2007). Ionic TMD–TMD interactions can be dynamic, such as in activation of voltage-activated ion channels. There, the sliding helix model posits that sequential formation of ion pairs between Arg residues of the S4 TMD with acidic residues of different surrounding TMDs stabilizes S4 in the membrane and permits its voltage-triggered movement (Zhang et al. 2007; DeCaen et al. 2008).

A set of high-affinity TMDs was recently isolated from a combinatorial library whose members contain both basic and acidic residues at certain positions (Herrmann et al. 2010). The invariant Leu-based host employed here is apparently hydrophobic enough to maintain polar residues within the membrane (Lew et al. 2000; Hessa et al. 2005). A detailed analysis of representative sequences indicated that ionic forces between appropriately spaced basic and acidic residues seem to be essential for interaction. Specifically, an ionizable residue at position i can interact with another one at position i – 1, i + 2, or i + 3. It is quite likely that additional productive combinations exist. Context dependence of these interfacial residues is again apparent since a C-terminal GxxxG starting at i + 7 is essential for high-affinity interaction and neighboring Ser, Cys, Tyr, or His residues contribute to the interfaces. Similar to the polar/His node discussed above, pre-orientation of the helices via interaction of the GxxxG motif may ensure precise geometrical positioning required for charge–charge interaction. Database searching yielded only few TMDs whose potential self-interaction is suggested by a pair of appropriately spaced ionizable residues in combination with GxxxG. However, hundreds of natural TMDs contain either a basic or an acidic residue plus GxxxG. The majority of the latter motifs was

175

overrepresented and might thus enter heterotypic interactions provided the spatio-temporal co-expression of the respective proteins (Herrmann et al. 2010). In addition to this cooperation of charged residues and GxxxG, it is clear that mechanisms not relying on GxxxG motifs must exist that allow for formation of salt-bridges between the TMDs of those natural proteins where experiment has clearly identified them, like the T-cell receptor.

3.5 Aromatic interactions

Evidence for aromatic interactions between TMDs was originally provided by the frequent interfacial positioning of Trp, Tyr, and Phe in polytopic proteins (Langosch and Heringa 1998; Adamian and Liang 2001; Adamian et al. 2003). Experimentally, a library screen showed that Trp residues of high-affinity TMDs prevailed at g positions of the randomized heptad motif. Mutation of Trp residues reduced self-interaction and grafting Trp residues onto artificial TMDs strongly enhanced their affinity (Ridder et al. 2005). A contribution of aromatic residues is also implied by the overabundance of WxxW and YxxY motifs in bacterial TMDs and mutational analysis of one candidate TMD that belongs to the cholera toxin secretion protein EpsM confirmed that WxxW, YxxW, WxxY, YxxY, and single Trp residues support its self-interaction (Sal-Man et al. 2007). A stabilizing role of aromatic–aromatic interactions was also seen when Phe, Tyr, and Trp promoted interaction of model TMDs. Further, cation–π interactions between aromatics and Arg, Lys, or His residues on the partner helix can lead to even higher TMD–TMD affinities (Johnson et al. 2007). In another study, a stabilizing role was observed for Phe when located at the i-3 position of GxxxG of high-affinity TMDs as isolated from a combinatorial library, thus yielding FxxGxxxG motifs (Unterreitmeier et al. 2007). This motif, and a number of analogs with different Phe/GxxxG spacings, is overrepresented in TMDs of natural bitopic membrane proteins. Within the framework of an oligo-Met host, only FxxGxxxG (present in >200 natural TMDs) self-interacted more strongly than GxxxG; thus, other overrepresented variants, such as FGxxxG, GxxFG, GxxxGF, GxxxGxF, GxxxGxxF, and GxxxGxxxF (>1300 natural TMDs) might support heterotypic interactions. It is currently not clear how Phe and GxxxG cooperate to form a helix–helix interface. The role of GxxxG might be to orient the Phe residues such as to promote aromatic–aromatic interactions. Alternatively, the Phe residue could interact with the backbone at a Gly of GxxxG of the partner helix via a C_α–H$\cdots\pi$ interaction known to be prevalent in soluble protein cores (Brandl et al. 2001). Albeit weak, these C_α–H$\cdots\pi$ interactions could be stabilized by the low dielectric environment of membranes as discussed above. A noteworthy observation is that the efficiencies by which the different aromatics stabilized TMD–TMD interactions in these studies vary widely. While Trp, but not Tyr, promoted interaction of a library isolate where GxxxG was absent (Ridder et al. 2005), self-interaction of certain model TMDs followed the order Phe>Tyr≈Trp (Johnson et al. 2007), and only Phe is effective at the –3 position of GxxxG (Unterreitmeier et al. 2007). Therefore, the mechanism and efficiency by which aromatics can induce helix–helix interactions seems to be strongly dependent on the surrounding structure.

4 Dynamic TMD–TMD interactions

TMD–TMD interfaces have been discussed above in the conceptual framework of static structures that are stabilized by mixtures of different non-covalent forces. This picture is appropriate in cases where TMDs mediate kinetically stable subunit oligomerization. It is clear, however, that these interactions may be reversible on time scales that are relevant for biological function. Thus, formation of a TMD–TMD interface appears to depend on the functional state of certain proteins, such as in signal transduction after ligand binding. TMDs may interact reversibly by translational movement within the bilayer plane, rotate relative to each others, or undergo even

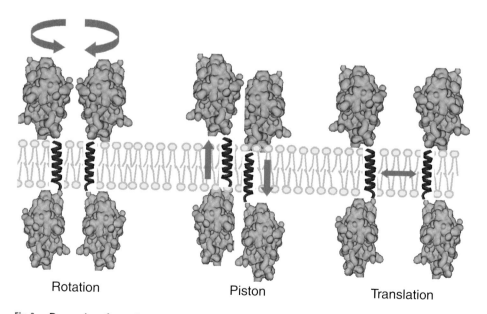

Rotation   Piston   Translation

Fig. 3. **Dynamics of membrane-embedded protein domains.** The activation of bitopic proteins upon binding of soluble ligands to extracellular domains has been proposed to involve the reorientation of transmembrane helices relative to each others about their long axes, reversible association/dissociation, and piston movements. Modified after this figure in Langosch and Arkin (2009).

piston motions (Fig. 3; Matthews et al. 2006; Moore et al. 2008). A few model proteins will be discussed here to illustrate the point. Reversible interactions involving translational movement are proposed to regulate the adhesive function of integrins (Gottschalk and Kessler 2002; Luo and Springer 2006). There, heterotypic TMD–TMD interactions between a set of α and β subunits (Gottschalk and Kessler 2004a,b; Schneider and Engelman 2004; Lin et al. 2006b; Lau et al. 2009) are thought to be displaced in favor of homotypic interaction (Li et al. 2004b) during activation (Li et al. 2004a). Rotation of TMDs relative to each other is a concept that appears to supersede the more traditional idea of ligand-induced dimerization of growth factor receptors. There is now substantial evidence that these receptors can exist as preformed dimers that are stabilized by TMD–TMD interactions. Receptor activation seems to involve TMD rotation in response to ligand-binding to extracellular domains, in case of erythropoietin (Seubert et al. 2003), epidermal growth factor (Moriki et al. 2001), and growth hormone (Brown et al. 2005) receptors. Interestingly, the arrangement of TMDs can also be influenced by direct binding of hydrophobic ligands. For example, the thrombopoietin receptor was activated by a synthetic compound that required a TMD His residue (Nakamura et al. 2006; Kim et al. 2007). Also, modeling studies suggest that the TMD of the ErbB2 tyrosine kinase is able to rotate to adopt two alternate dimerization motifs, thereby controlling the activity of the protein (Fleishman et al. 2002). This view is largely compatible with the idea that upon receptor activation helix–helix interaction moves from the N-terminal helix half to the C-terminal half (Escher et al. 2009). Changing the electrostatics between TMDs is another way to change their orientation relative to each other. The homotetrameric M2 protein from influenza A forms a proton channel, which is activated by lowering the pH. Its TM-helices cross each other at positive angles as indicated by earlier functional (Pinto et al. 1997), biochemical (Bauer et al. 1999), and modeling (Dieckmann and DeGrado 1997) work. The high-resolution structures which have been solved recently (Schnell and Chou 2008; Stouffer et al. 2008) suggest that His protonation promotes channel gating, although it still remains unknown how exactly a pH change opens the pore. Linear and 2D-IR spectroscopic studies have provided evidence that is consistent with a rotation of the helices about their long axes upon pH change (Manor et al. 2009). This rotational change is on the order of one amino acid register and may provide a molecular picture of channel gating.

Acknowledgments

I thank all of my past and present co-workers for their dedicated contributions to research in my lab. Also, I apologize to the many researchers whose contributions to the field could not be included here due to space constraints. Work in the author's laboratory is supported by the Deutsche Forschungsgemeinschaft, the Munich Center for Integrated Protein Science, CIPSM, the BMBF, and the State of Bavaria.

References

Adamian L and Liang J (2001) Helix–helix packing and interfacial pairwise interactions of residues in membrane proteins. J Mol Biol 311: 891–907

Adamian L and Liang J (2002) Interhelical hydrogen bonds and spatial motifs in membrane proteins: polar clamps and serine zippers. Proteins Struct Funct Genet 47: 209–218

Adamian L, Jackups R Jr, Binkowski TA, Liang J (2003) Higher-order interhelical spatial interactions in membrane proteins. J Mol Biol 327: 251–272

Adams PD, Engelman DM, Brünger AT (1996) Improved prediction for the structure of the dimeric transmembrane domain of glycophorin A obtained through global searching. Proteins 26: 257–261

Arbely E and Arkin IT (2004) Experimental measurement of the strength of alpha Ca–H...O bond in a lipid bilayer. J Am Chem Soc 126: 5362–5363

Arbely E, Granot Z, Kass I, Orly J, Arkin IT (2006) A trimerizing GxxxG motif is uniquely inserted in the severe acute respiratory syndrome (SARS) coronavirus spike protein transmembrane domain. Biochemistry 45: 11349–11356

Arkin IT (2002) Structural aspects of oligomerization taking place between the transmembrane alpha-helices of bitopic membrane proteins. Biochim Biophys Acta 1565: 347–363

Arkin IT and Brünger AT (1998) Statistical analysis of predicted transmembrane alpha-helices. Biochim Biophys Acta 1429: 113–128

Ash WL, Stockner T, MacCallum JL, Tieleman DP (2004) Computer modeling of polyleucine-based coiled coil dimers in a realistic membrane environment: insight into helix–helix interactions in membrane proteins. Biochemistry 43: 9050–9060

Asundi VK and Carey DJ (1995) Self-association of N-syndecan (syndecan-3) core protein is mediated by a novel structural motif in the transmembrane domain and ectodomain flanking region. J Biol Chem 270: 26404–26410

Barwe SP, Kim S, Rajasekaran SA, Bowie JU, Rajasekaran AK (2007) Janus model of the Na, K-ATPase beta-subunit transmembrane domain: distinct faces mediate alpha/beta assembly and beta–beta homo-oligomerization. J Mol Biol 365: 706–714

Bauer CM, Pinto LH, Cross TA, Lamb RA (1999) The influenza virus M2 ion channel protein: probing the structure of the transmembrane domain in intact cells by using engineered disulfide cross-linking. Virology 254: 196–209

Beel AJ, Mobley CK, Kim HJ, Tian F, Hadziselimovic A, Jap B, Prestegard JH, Sanders CR (2008) Structural studies of the transmembrane C-terminal domain of the amyloid precursor protein (APP): does APP function as a cholesterol sensor? Biochemistry 47: 9428–9446

Bocharov EV, Pustovalova YE, Pavlov KV, Volynsky PE, Goncharuk MV, Ermolyuk YS, Karpunin DV, Schulga AA, Kirpichnikov MP, Efremov RG, Maslennikov IV, Arseniev AS (2007) Unique dimeric structure of BNip3 transmembrane domain suggests membrane permeabilization as a cell death trigger. J Biol Chem 282: 16256–16266

Bocharov EV, Mayzel ML, Volynsky PE, Goncharuk MV, Ermolyuk YS, Schulga AA, Artemenko EO, Efremov RG, Arseniev AS (2008a) Spatial structure and pH-dependent conformational diversity of dimeric transmembrane domain of the receptor tyrosine kinase EphA1. J Biol Chem 283: 29385–29395

179

Bocharov EV, Mineev KS, Volynsky PE, Ermolyuk YS, Tkach EN, Sobol AG, Chupin VV, Kirpichnikov MP, Efremov RG, Arseniev AS (2008b) Spatial structure of the dimeric transmembrane domain of the growth factor receptor ErbB2 presumably corresponding to the receptor active state. J Biol Chem 283: 6950–6956

Bormann B-J, Knowles WJ, Marchesi VT (1989) Synthetic peptides mimic the assembly of transmembrane glycoproteins. J Biol Chem 264: 4033–4037

Bowie JU (1997) Helix packing in membrane proteins. J Mol Biol 272: 780–789

Bowie JU (2005) Solving the membrane protein folding problem. Nature 438: 581–589

Brandl M, Weiss MS, Jabs A, Sühnel J, Hilgenfeld R (2001) C–H...π interactions in proteins. J Mol Biol 307: 357–377

Brosig B and Langosch D (1998) The dimerization motif of the glycophorin A transmembrane segment in membranes: importance of glycine residues. Protein Sci 7: 1052–1056

Brown RJ, Adams JJ, Pelekanos RA, Wan Y, McKinstry WJ, Palethorpe K, Seeber RM, Monks TA, Eidne KA, Parker MW, Waters MJ (2005) Model for growth hormone receptor activation based on subunit rotation within a receptor dimer. Nat Struct Mol Biol 12: 814–821

Call ME, Schnell JR, Xu CQ, Lutz RA, Chou JJ, Wucherpfennig KW (2006) The structure of the zeta transmembrane dimer reveals features essential for its assembly with the T cell receptor. Cell 127: 355–368

Call ME and Wucherpfennig KW (2007) Common themes in the assembly and architecture of activating immune receptors. Nat Rev Immunol 7: 841–850

Chamberlain AK, Faham S, Yohannan S, Bowie JU (2003) Construction of helix-bundle membrane proteins. Adv Protein Chem 63: 19–46

Chin CN and von Heijne G (2000) Charge pair interactions in a model transmembrane helix in the ER membrane. J Mol Biol 303: 1–5

Chin CN, Sachs JN, Engelman DM (2005) Transmembrane homodimerization of receptor-like protein tyrosine phosphatases. FEBS Lett 579: 3855–3858

Choma C, Gratkowski H, Lear JD, DeGrado WF (2000) Asparagine-mediated self-association of a model transmembrane helix. Nature Struct Biol 7: 161–166

Chothia C (1984) Principles that determine the structure of proteins. Annu Rev Biochem 53: 537–572

Dawson JP, Weinger JS, Engelman DM (2002) Motifs of serine and threonine can drive association of transmembrane helices. J Mol Biol 316: 799–805

DeCaen PG, Yarov-Yarovoy V, Zhao Y, Scheuer T, Catterall WA (2008) Disulfide locking a sodium channel voltage sensor reveals ion pair formation during activation. Proc Natl Acad Sci USA 105: 15142–15147

DeGrado WF, Gratkowski H, Lear JD (2003) How do helix–helix interactions help determine the folds of membrane proteins? Perspectives from the study of homo-oligomeric helical bundles. Prot Sci 12: 647–665

Dews IC, MacKenzie KR (2007) Transmembrane domains of the syndecan family of growth factor coreceptors display a hierarchy of homotypic and heterotypic interactions. Proc Natl Acad Sci USA 104: 20782–20787

Dieckmann GR and DeGrado WF (1997) Modeling transmembrane helical oligomers. Curr Opin Struct Biol 7: 486–494

Doura AK and Fleming KG (2004) Complex interactions at the helix–helix interface stabilize the glycophorin A transmembrane dimer. J Mol Biol 343: 1487–1497

Doura AK, Kobus FJ, Dubrovsky L, Hibbard E, Fleming KG (2004) Sequence context modulates the stability of a GxxxG-mediated transmembrane helix–helix Dimer. J Mol Biol 341: 991–998

experimentally and computationally. Thermodynamic measurements (Fleming and Engelman 2001; Cristian et al. 2003) as well as genetic approaches (Langosch et al. 1996; Russ and Engelman 1999) have been developed and applied to determine the strength of individual helix interactions and estimate the effect of mutations on helix assembly (for a recent review see MacKenzie and Fleming 2008). Available membrane protein structures have been rigorously analyzed to obtain insights into helix–helix packing and to derive amino acid propensities for the participation in interhelical interactions (Adamian and Liang 2001; Eilers et al. 2002; Gimpelev et al. 2004).

2.1 Diversity of helix–helix contacts in membrane proteins

When analyzing the nature of non-covalent interactions within membrane proteins, polar residues as well as small residues have been found to strongly contribute to the transmembrane helix assembly despite the overall hydrophobic nature of transmembrane regions and the accordingly large contribution of Van der Waals interactions. In fact, several groups observed that the composition of helix interfaces in membrane proteins is even more diverse than in soluble proteins (Adamian and Liang 2001; Eilers et al. 2002). While soluble proteins have a strong preference for salt-bridge interactions formed by oppositely ionizable amino acids, membrane proteins feature a much broader range of polar interactions involving residue pairs formed by different polar residues such as S, T, Y, N, and Q (Adamian and Liang 2001). On average, every transmembrane helix is expected to form at least one hydrogen bond with another helix. Especially side-chain backbone hydrogen bonds contribute substantially to this observation as every second hydrogen bond between transmembrane helices seems to be of this type (Gimpelev et al. 2004). Experimentally, transmembrane helices with motifs of multiple serine and threonine residues or single glutamine, asparagine, aspartic acid or glutamic acid residues were found to promote strong self interaction (Dawson et al. 2002; Zhou et al. 2000), further reinforcing the importance of polar residues for helix association.

Small residues (G, A, and S), on the other hand, were repeatedly found to be among the most over-represented residues within membrane helix interaction interfaces (Eilers et al. 2002; Gimpelev et al. 2004). These residues found even more attention after it had been reported that motifs consisting of two small residues spaced by three residues ([GAS]xxx[GAS]) are a recurrent theme in helix–helix interfaces (Lemmon et al. 1992; Brosig and Langosch 1998; Mendrola et al. 2002; Overton et al. 2003; Lee et al. 2004). Generally, small residues are thought to allow very close contact between transmembrane helices and accordingly extensive van der Waals interactions (Javadpour et al. 1999) as well as the formation of $C\alpha–H\cdots O$ hydrogen bonds across the helical backbone (Senes et al. 2001). Lately, however, measurements of helix interaction energies have indicated that GxxxG-containing

189

transmembrane segments may interact with remarkably different strength suggesting that sequence context is equally important for interaction as the GxxxG motif itself (Doura et al. 2004; MacKenzie and Fleming 2008).

2.2 Frequency of residue contacts in membrane and soluble proteins

It is well known that the prediction of intra-molecular amino acid contacts becomes more difficult with decreasing contact density (fraction of observed contacts among the total number of possible residue pairs; Punta and Rost 2005). For soluble proteins, the contact density was found to depend strongly on the secondary structure content of a protein, with all-alpha proteins having roughly only half as many contacts as all-beta proteins (Punta and Rost 2005). This finding explains repeatedly reported difficulties in contact prediction of all-alpha proteins.

Comparing the frequency of residue contacts in membrane and soluble proteins, Fig. 1 shows the dependency of the number of observed contacts in a protein on protein length for three different types of proteins: soluble proteins, soluble proteins in the SCOP class all-alpha, and alpha-helical membrane proteins (only transmembrane segments considered). While all-alpha soluble proteins are found to possess slightly fewer contacts than soluble proteins in general over the

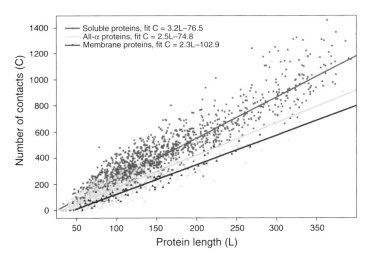

Fig. 1. Contact density (number of contacts as a function of protein length) of membrane proteins compared to soluble proteins. The amount of contacts for any type of protein is linearly proportional to protein length, with membrane proteins having generally fewer contacts than soluble proteins. The fitted curves represent contact functions expressing the expected number of contacts for a given protein class and length. Figure adapted from Fuchs et al. (2009).

190

full range of analyzed protein lengths, the number of observed contacts within membrane proteins is even more reduced compared to soluble proteins in general and all-alpha soluble proteins in particular. Accordingly, the prediction of helix–helix contacts in membrane proteins can be expected to be at least of comparable difficulty to the prediction of intra-molecular contacts within all-alpha soluble proteins, if not more difficult.

3 Prediction of lipid accessibility

Numerous approaches have been developed for the prediction of solvent accessibility in soluble proteins (for examples, see Rost and Sander 1994; Pascarella et al. 1998; Li and Pan 2001; Pollastri et al. 2002; Yuan et al. 2002; Wang et al. 2005). Similarly, lipid accessibility of transmembrane residues is an important prediction task in the field of membrane protein structural bioinformatics. According to the two-stage model of membrane protein folding, transmembrane helices are formed and inserted into the membrane at the first folding stage and subsequently associate into the final structure during the second folding stage (Popot and Engelman 1990). While topology prediction programs address the first stage by detecting transmembrane helices based on hydrophobicity, lipid accessibility predictions may help to reproduce the second stage as they provide information about the likely orientation of individual transmembrane helices toward the membrane, thereby indirectly identifying the position of potential helix interaction interfaces.

In fact, when membrane proteins were still believed to fold into mostly canonical helix bundles where adjacent helices are oriented to one another in either parallel or anti-parallel fashion, the correctly predicted orientation of transmembrane helices seemed to be sufficient for the accurate modeling of alpha-helical membrane proteins. Historically, the development of lipid accessibility methods has therefore found much more attention than the prediction of individual residue contacts within transmembrane portions and the number of methods and publicly available web servers in this field is still growing (see Helms et al., this volume, for a detailed description of these methods).

Historically, three major strategies for predicting lipid accessibility were proposed which will be briefly summarized in the following paragraphs. Recent methods tend to combine the most successful of these strategies leading to significantly improved prediction accuracies as described at the end of this section.

3.1 Hydrophobicity-based predictions

Soluble proteins are well known to have a distinct distribution between exposed polar residues and hydrophobic buried residues. Membrane proteins were similarly suspected to follow an inversed hydrophobic pattern with strongly hydrophobic residues facing the membrane environment and less hydrophobic residues buried inside the helix bundle. Accordingly, early approaches for predicting the angular orientation of transmembrane helices used hydrophobicity scales to derive the hydrophobic or hydrophilic moment of a helix which was assumed to indicate its lipid exposed or buried face (Rees et al. 1989, for a mathematical definition of the hydrophobic moment see also Helms et al., this volume).

However, an increasing number of available membrane protein structures have meanwhile unambiguously demonstrated that the hydrophobic organization of membrane proteins is not following a clear "inside-out" pattern (Stevens and Arkin 1999; Mokrab et al. 2009) with only small differences in hydrophobicity being observed between the surface and interior of transmembrane proteins. Hydrophobicity is therefore only weakly correlated with the lipid exposure of a transmembrane helix and is not used in recent prediction methods.

3.2 Amino acid propensity scales derived from membrane protein sequences and structures

While membrane protein structures were still too scarce to allow for statistically solid analyses of amino acid frequencies in exposed and buried positions, several approaches were developed to derive lipid exposure propensities for individual amino acids from membrane protein sequences only. The most prominent example is the kPROT scale which compares amino acid frequencies in multi-span membrane proteins to those in single-span proteins assuming that these two classes represent an approximation to amino acid distributions in buried and exposed positions, respectively (Pilpel et al. 1999). While kPROT propensities could be shown to be in good agreement with another sequence-based scale introduced by Samatey et al. (1995) and were able to predict membrane facing vectors with better accuracy than any other method available at that time, recent analyses of membrane protein structures have indicated that sequence-derived propensity scales may deviate remarkably from observed amino acid distributions.

With increasing numbers of available membrane protein structures, several analyses have compared the composition of membrane protein surface and buried areas, sometimes also depending on a residue's position with respect to the membrane (Beuming and Weinstein 2004; Adamian et al. 2005; Mokrab et al. 2009) or considering different functional classes of membrane proteins (Hildebrand et al. 2006). Consistently, these studies reported a strongly hydrophobic composition of surfaces situated within the membrane core while small side-chains have a strong tendency to be enriched in buried positions. Nevertheless, aromatic and charged amino acids may be found in lipid-exposed positions as well, albeit more often situated in terminal regions of the membrane than in the hydrocarbon core. Importantly, the interior of transmembrane helix bundles is still quite hydrophobic with a similar average hydrophobicity as found within the interior of soluble proteins.

Several propensity scales have been derived either from amino acid counts in exposed and buried positions (Adamian et al. 2005; Mokrab et al. 2009), from surface fractions (Beuming and Weinstein 2004) or from relative solvent-accessible surface area (rSASA) values (Park and Helms 2007). Generally, structure-derived propensity scales have been shown to outperform sequence-based scales (Park and Helms 2006b) and are now commonly incorporated into recent lipid accessibility predictors (Beuming and Weinstein 2004; Adamian and Liang 2006; Hildebrand et al. 2006). The derivation of lipid exposure propensity scales is discussed in detail by Helms et al. (this volume).

3.3 Sequence conservation of exposed and buried transmembrane residues

Consistent with the properties of soluble proteins, surface residues in membrane proteins are known to be less conserved than buried positions. Accordingly, sequence conservation is an effective feature for discriminating lipid exposed from interior residues and has found widespread application in the prediction of lipid accessibility (Rees et al. 1989; Donnelly et al. 1993; Beuming and Weinstein 2004; Adamian and Liang 2006; Park et al. 2007). In fact, a comparative analysis has shown that sequence conservation is more strongly correlated with exposure patterns of transmembrane helices than available empirical propensity and hydrophobicity scales (Park and Helms 2006b) and hence is so far the most informative sequence feature related to residue exposure in membrane proteins.

3.4 Best performing methods in the field of lipid accessibility

All state-of-the-art methods for the prediction of exposed and buried transmembrane residues or helix faces rely on sequence conservation in combination with structure-derived propensity scales. Various approaches have been proposed to combine propensity values with measures of sequence conservation into a general score. Individual residues are then predicted as buried or exposed either using a predefined score threshold (Beuming and Weinstein 2004; Rose et al. 2009), or with the assistance of a support vector classifier (Park et al. 2007). Alternatively, helix–lipid and helix–helix interfaces are identified by sorting all possible helix faces according to their calculated average score (Adamian and Liang 2006).

Currently the best prediction accuracy for the prediction of buried and exposed residues is reported for the TMX method, where 79% of all residues were correctly predicted given a dataset of 41 membrane protein chains (Park et al. 2007). Complete helix faces are best predicted using the LIPS method introduced by Adamian and Liang (2006) which identified helix–lipid interfaces from 18 membrane proteins with an accuracy of 88% while the most buried helix face was correctly identified in 70% of all cases.

4 Prediction of helix–helix contacts

While the prediction of lipid accessibility provides crucial hints for identifying individual residues likely to participate in helix–helix contacts, methods dealing specifically with the prediction of residue pairs forming a helix–helix contact were completely lacking until just recently. In contrast, contact prediction in soluble proteins has a long history, from simple prediction methods focusing on residue co-evolution (Gobel et al. 1994; Olmea and Valencia 1997) to complex prediction algorithms using machine-learning techniques to evaluate a large number of different sequence features (Fariselli et al. 2001b; Punta and Rost 2005; Cheng and Baldi 2007).

Recent membrane protein structures have been found to deviate remarkably from canonical helix bundles, with helices not necessarily oriented in parallel or anti-parallel fashion with respect to each other and many interactions taking place between sequentially distant helices (Elofsson and von Heijne 2007). This observation has motivated now the development of contact prediction methods specifically for membrane proteins since additional constraints are needed to model such complex helix bundles. Furthermore, contact prediction methods for soluble proteins have been shown to perform only poorly on membrane proteins due to their distinct amino acid composition (Fuchs et al. 2009). Following the historical course of contact prediction in soluble proteins, the subsequent paragraphs will first summarize a study on co-evolving residues and their relationship to helix–helix contacts in membrane proteins and will then present and compare two methods using different machine-learning techniques for the prediction of residue contacts between transmembrane helices.

4.1 Co-evolving residues in membrane proteins

Mutations that tend to destabilize a particular protein structure may provoke other positions to mutate concurrently in order to compensate for the loss of stability. Amino acid contacts have been suggested to be primary spots of these compensatory processes, making the detection of sequence positions with correlated mutational behavior an important feature for residue contact prediction methods (Gobel et al. 1994). Recently, a first analysis of correlated mutational behavior within membrane

Table 1. Contact prediction accuracies obtained by different methods for a non-redundant dataset of 14 membrane protein structures

Method	Reference	Contact prediction accuracy (%)	Prediction accuracy ($|\delta| = 4$) (%)
HelixCorr	Fuchs et al. (2007)	12.0	55.0
TMHcon	Fuchs et al. (2009)	31.8	79.3
TMhit	Lo et al. (2009)	31.0	56.8

Only the best L/5 predicted contacts were considered, where L is the total number of transmembrane residues. Contact prediction accuracy: fraction of correctly predicted contacts out of all predicted contacts. Prediction accuracy ($|\delta| = 4$): fraction of predicted contacts lying within one helix turn of an observed contact.

proteins was conducted (Fuchs et al. 2007) which had been impossible before due to the paucity of available membrane protein structures.

In agreement with studies conducted on soluble proteins, it could be observed that only a small fraction of co-evolving residues actually involved helix–helix contact pairs. However, up to 50% of all strongly correlated residue pairs detected with different prediction methods were found to be in close vicinity to interhelical contacts. Combining the outcome of several prediction methods into a consensus prediction (termed HelixCorr), this fraction could be further increased to more than 55%. Recent publications analyzing co-evolving residues already highlighted that residue co-evolution may have other than structural reasons (Gloor et al. 2005; Lee et al. 2008). The results obtained with HelixCorr additionally indicated that co-evolution not only occurs to maintain specific amino acids required for a structural contact but also influences the correct formation of a helix–helix contact by affecting its sequence context.

Although prediction accuracies obtained in the study of Fuchs et al. (2007) were clearly too low to make co-evolving residues alone a useful prediction method for helix–helix contacts (Table 1), their frequent occurrence in close sequence neighborhood to real helix–helix contacts suggested that they might be an important source of information to derive helix pairs that are likely to be in direct contact, possibly along with the approximate region of interaction. Furthermore, given that prediction accuracies were largely consistent with those reported for soluble proteins, the combination of residue co-evolution with other sequence features promised further gain in prediction accuracy as previously demonstrated for soluble proteins (Olmea and Valencia 1997; Fariselli et al. 2001a).

4.2 Prediction of helix–helix contacts with machine-learning techniques

When contact prediction approaches for soluble proteins were still mainly focused on the analysis of co-evolving residues, additional sequence information was already

195

shown to improve the accuracy of obtained predictions significantly (Olmea and Valencia 1997). Accordingly, machine-learning methods which are able to incorporate a variety of sequence features have been consistently demonstrated to outperform methods using co-evolving residues alone (Fariselli et al. 2001a,b). Generally, these methods require contact maps of proteins with known structures for training. During the training phase, the machine-learning algorithm tries to deduce association rules between selected sequence features of each protein in the training set and its contact map. These rules are then applied to proteins without known contact map.

Over the years, several different implementations of machine-learning approaches have been applied for the residue contact prediction problem in soluble proteins with neural networks (Fariselli et al. 2001a,b; Punta and Rost 2005) and support vector machines (Cheng and Baldi 2007) being the most commonly used ones. For membrane proteins, there are now two machine-learning-based contact predictors available, the neural network-based predictor TMHcon (Fuchs et al. 2009) and the support vector machine-based predictor TMhit (Lo et al. 2009).

Both methods are specific for alpha-helical membrane proteins due to two reasons. Firstly, they are trained on datasets consisting solely of membrane proteins with solved structure and hence are adjusted to the hydrophobic nature of transmembrane segments and the specific properties of helix–helix contacts between these segments. Secondly, they combine sequence features known to be informative for contact prediction from soluble proteins with specific features that can only be derived for membrane proteins with alpha-helix bundle fold. Such membrane protein-specific features include relative residue positioning along the membrane (e.g., whether a residue is found close to the intra- or extracellular side or the core of the membrane), transmembrane helix lengths, and predicted lipid accessibility values. Significantly, for both methods these membrane protein-specific features made a particularly important contribution for improving prediction accuracy.

Tested on a non-redundant dataset of 14 membrane protein structures (listed in Supplementary Table 2 of Lo et al. 2009), both methods resulted in basically identical prediction accuracies of 31% (e.g., 31% of all predicted contacts are actual helix–helix contacts; Table 1). Compared to a contact prediction based only on co-evolving residues as implemented in the method HelixCorr, the application of machine-learning methods resulted in the improvement of prediction accuracies by nearly 20%. Furthermore, both methods are able to predict contacts in membrane proteins with at least equal accuracies as reported for comparable methods available for soluble proteins (for a comparative assessment of soluble contact predictors, see for example, Grana et al. 2005).

Interestingly, best results were consistently obtained by both TMHcon and TMhit for the important class of proteins with seven transmembrane helices indicating that

196

Acknowledgement

We would like to thank Jennifer Liss for careful reading of the manuscript and useful corrections.

References

Adamian L and Liang J (2001) Helix–helix packing and interfacial pairwise interactions of residues in membrane proteins. J Mol Biol 311: 891–907

Adamian L and Liang J (2006) Prediction of transmembrane helix orientation in polytopic membrane proteins. BMC Struct Biol 6: 13

Adamian L, Nanda V, DeGrado WF, Liang J (2005) Empirical lipid propensities of amino acid residues in multispan alpha helical membrane proteins. Proteins 59: 496–509

Barth P, Wallner B, Baker D (2009) Prediction of membrane protein structures with complex topologies using limited constraints. Proc Natl Acad Sci USA 106: 1409–1414

Beuming T and Weinstein H (2004) A knowledge-based scale for the analysis and prediction of buried and exposed faces of transmembrane domain proteins. Bioinformatics 20: 1822–1835

Brosig B and Langosch D (1998) The dimerization motif of the glycophorin A transmembrane segment in membranes: importance of glycine residues. Protein Sci 7: 1052–1056

Cheng J and Baldi P (2007) Improved residue contact prediction using support vector machines and a large feature set. BMC Bioinform 8: 113

Cristian L, Lear JD, DeGrado WF (2003) Use of thiol-disulfide equilibria to measure the energetics of assembly of transmembrane helices in phospholipid bilayers. Proc Natl Acad Sci USA 100: 14772–14777

Dawson JP, Weinger JS, Engelman DM (2002) Motifs of serine and threonine can drive association of transmembrane helices. J Mol Biol 316: 799–805

Donnelly D, Overington JP, Ruffle SV, Nugent JH, Blundell TL (1993) Modeling alpha-helical transmembrane domains: the calculation and use of substitution tables for lipid-facing residues. Protein Sci 2: 55–70

Doura AK, Kobus FJ, Dubrovsky L, Hibbard E, Fleming KG (2004) Sequence context modulates the stability of a GxxxG-mediated transmembrane helix–helix dimer. J Mol Biol 341: 991–998

Eilers M, Patel AB, Liu W, Smith SO (2002) Comparison of helix interactions in membrane and soluble alpha-bundle proteins. Biophys J 82: 2720–2736

Elofsson A and von Heijne G (2007) Membrane protein structure: prediction versus reality. Annu Rev Biochem 76: 125–140

Fariselli P, Olmea O, Valencia A, Casadio R (2001a) Prediction of contact maps with neural networks and correlated mutations. Protein Eng 14: 835–843

Fariselli P, Olmea O, Valencia A, Casadio R (2001b) Progress in predicting inter-residue contacts of proteins with neural networks and correlated mutations. Proteins Suppl 5: 157–162

Fleishman SJ, Unger VM, Yeager M, Ben-Tal N (2004a) A Calpha model for the transmembrane alpha helices of gap junction intercellular channels. Mol Cell 15: 879–888

Fleishman SJ, Yifrach O, Ben-Tal N (2004b) An evolutionarily conserved network of amino acids mediates gating in voltage-dependent potassium channels. J Mol Biol 340: 307–318

Fleishman SJ, Unger VM, Ben-Tal N (2006) Transmembrane protein structures without X-rays. Trends Biochem Sci 31: 106–113

Fleming KG and Engelman DM (2001) Specificity in transmembrane helix–helix interactions can define a hierarchy of stability for sequence variants. Proc Natl Acad Sci USA 98: 14340–14344

Fuchs A, Martin-Galiano AJ, Kalman M, Fleishman S, Ben-Tal N, Frishman D (2007) Co-evolving residues in membrane proteins. Bioinformatics 23: 3312–3319

Fuchs A, Kirschner A, Frishman D (2009) Prediction of helix–helix contacts and interacting helices in polytopic membrane proteins using neural networks. Proteins 74: 857–871

Gimpelev M, Forrest LR, Murray D, Honig B (2004) Helical packing patterns in membrane and soluble proteins. Biophys J 87: 4075–4086

Gloor GB, Martin LC, Wahl LM, Dunn SD (2005) Mutual information in protein multiple sequence alignments reveals two classes of coevolving positions. Biochemistry 44: 7156–7165

Gobel U, Sander C, Schneider R, Valencia A (1994) Correlated mutations and residue contacts in proteins. Proteins 18: 309–317

Grana O, Baker D, MacCallum RM, Meiler J, Punta M, Rost B, Tress ML, Valencia A (2005) CASP6 assessment of contact prediction. Proteins 61 (Suppl 7): 214–224

Hildebrand PW, Lorenzen S, Goede A, Preissner R (2006) Analysis and prediction of helix–helix interactions in membrane channels and transporters. Proteins 64: 253–262

Javadpour MM, Eilers M, Groesbeek M, Smith SO (1999) Helix packing in polytopic membrane proteins: role of glycine in transmembrane helix association. Biophys J 77: 1609–1618

Langosch D, Brosig B, Kolmar H, Fritz HJ (1996) Dimerisation of the glycophorin A transmembrane segment in membranes probed with the ToxR transcription activator. J Mol Biol 263: 525–530

Lee BC, Park K, Kim D (2008) Analysis of the residue-residue coevolution network and the functionally important residues in proteins. Proteins 72: 863–872

Lee SF, Shah S, Yu C, Wigley WC, Li H, Lim M, Pedersen K, Han W, Thomas P, Lundkvist J, Hao YH, Yu G (2004) A conserved GXXXG motif in APH-1 is critical for assembly and activity of the gamma-secretase complex. J Biol Chem 279: 4144–4152

Lemmon MA, Flanagan JM, Treutlein HR, Zhang J, Engelman DM (1992) Sequence specificity in the dimerization of transmembrane alpha-helices. Biochemistry 31: 12719–12725

Li X and Pan XM (2001) New method for accurate prediction of solvent accessibility from protein sequence. Proteins 42: 1–5

Lo A, Chiu YY, Rodland EA, Lyu PC, Sung TY, Hsu WL (2009) Predicting helix–helix interactions from residue contacts in membrane proteins. Bioinformatics 25: 996–1003

Lundstrom K (2005) Structural genomics of GPCRs. Trends Biotechnol 23: 103–108

MacKenzie KR and Fleming KG (2008) Association energetics of membrane spanning alpha-helices. Curr Opin Struct Biol 18: 412–419

Mendrola JM, Berger MB, King MC, Lemmon MA (2002) The single transmembrane domains of ErbB receptors self-associate in cell membranes. J Biol Chem 277: 4704–4712

Mokrab Y, Stevens TJ, Mizuguchi K (2009) Lipophobicity and the residue environments of the transmembrane alpha-helical bundle. Proteins 74: 32–49

Olmea O and Valencia A (1997) Improving contact predictions by the combination of correlated mutations and other sources of sequence information. Fold Des 2: S25–S32

Overton MC, Chinault SL, Blumer KJ (2003) Oligomerization, biogenesis, and signaling is promoted by a glycophorin A-like dimerization motif in transmembrane domain 1 of a yeast G protein-coupled receptor. J Biol Chem 278: 49369–49377

Park Y and Helms V (2006a) Assembly of transmembrane helices of simple polytopic membrane proteins from sequence conservation patterns. Proteins 64: 895–905

Park Y and Helms V (2006b) How strongly do sequence conservation patterns and empirical scales correlate with exposure patterns of transmembrane helices of membrane proteins? Biopolymers 83: 389–399

Park Y and Helms V (2007) On the derivation of propensity scales for predicting exposed transmembrane residues of helical membrane proteins. Bioinformatics 23: 701–708

Park Y, Hayat S, Helms V (2007) Prediction of the burial status of transmembrane residues of helical membrane proteins. BMC Bioinform 8: 302

Pascarella S, De Persio R, Bossa F, Argos P (1998) Easy method to predict solvent accessibility from multiple protein sequence alignments. Proteins 32: 190–199

Pilpel Y, Ben-Tal N, Lancet D (1999) kPROT: a knowledge-based scale for the propensity of residue orientation in transmembrane segments. Application to membrane protein structure prediction. J Mol Biol 294: 921–935

Pollastri G, Baldi P, Fariselli P, Casadio R (2002) Prediction of coordination number and relative solvent accessibility in proteins. Proteins 47: 142–153

Popot JL and Engelman DM (1990) Membrane protein folding and oligomerization: the two-stage model. Biochemistry 29: 4031–4037

Punta M and Rost B (2005) PROFcon: novel prediction of long-range contacts. Bioinformatics 21: 2960–2968

Rees DC, DeAntonio L, Eisenberg D (1989) Hydrophobic organization of membrane proteins. Science 245: 510–513

Rose A, Lorenzen S, Goede A, Gruening B, Hildebrand PW (2009) RHYTHM–a server to predict the orientation of transmembrane helices in channels and membrane-coils. Nucleic Acids Res 37: W575–W580

Rost B and Sander C (1994) Conservation and prediction of solvent accessibility in protein families. Proteins 20: 216–226

Russ WP and Engelman DM (1999) TOXCAT: a measure of transmembrane helix association in a biological membrane. Proc Natl Acad Sci USA 96: 863–868

Samatey FA, Xu C, Popot JL (1995) On the distribution of amino acid residues in transmembrane alpha-helix bundles. Proc Natl Acad Sci USA 92: 4577–4581

Senes A, Ubarretxena-Belandia I, Engelman DM (2001) The Calpha–H⋯O hydrogen bond: a determinant of stability and specificity in transmembrane helix interactions. Proc Natl Acad Sci USA 98: 9056–9061

Stevens TJ and Arkin IT (1999) Are membrane proteins "inside-out" proteins? Proteins 36: 135–143

Taylor WR, Jones DT, Green NM (1994) A method for alpha-helical integral membrane protein fold prediction. Proteins 18: 281–294

Wang JY, Lee HM, Ahmad S (2005) Prediction and evolutionary information analysis of protein solvent accessibility using multiple linear regression. Proteins 61: 481–491

Yuan Z, Burrage K, Mattick JS (2002) Prediction of protein solvent accessibility using support vector machines. Proteins 48: 566–570

Zhou FX, Cocco MJ, Russ WP, Brunger AT, Engelman DM (2000) Interhelical hydrogen bonding drives strong interactions in membrane proteins. Nat Struct Biol 7: 154–160

Natural constraints, folding, motion, and structural stability in transmembrane helical proteins

Susan E. Harrington and Nir Ben-Tal

Department of Biochemistry and Molecular Biology, George S. Wise Faculty of Life Sciences, Tel Aviv University, Ramat Aviv, Israel

Abstract

Transmembrane (TM) helical proteins are of fundamental importance in many diverse biological processes. To understand these proteins functionally, it is necessary to characterize the forces that stabilize them. What are these forces (both within the protein itself and between the protein and membrane) and how do they give rise to the multiple conformational states and complex activity of TM helical proteins? How do they act in concert to fold TM helical proteins, create their low-energy stable states, and guide their motion? These central questions have led to the description of critical natural constraints and partial answers, which we will review. We will then describe how these constraints can be tracked through homologs and proteins of similar folds in order to better understand how amino acid sequence can specify structure and guide motion. Our emphasis throughout will be on structural features of TM helix bundles themselves, but we will also sketch the membrane-related aspects of these questions.

1 Folding background

Central aspects of transmembrane (TM) helical protein folding are well understood and have important structural implications (Bowie 2005).

1.1 Two-stage hypothesis

The widely accepted two-stage hypothesis provides the foundation for much current work in TM helical protein folding and structure prediction. It states that TM

Corresponding author: Susan E. Harrington, Department of Biochemistry and Molecular Biology, George S. Wise Faculty of Life Sciences, Tel Aviv University, Ramat Aviv 69978, Israel (E-mail: susan.e.harrington@gmail.com)

helices fold autonomously in the first stage of folding within the membrane, then associate to form helix pairs, then triples or quadruples and so forth, eventually building up to the full bundle (Popot and Engelman 1990). The two-stage hypothesis is known (and was known by its creators) to be a simplification, but it is a very useful one. It has led to the fruitful investigation of the association of helix pairs as the foundation for the folding of the full bundle and the structure prediction of TM helical proteins.

1.2 Translocon-aided folding

As they come off the ribosome during translation, TM helical proteins enter the membrane (the endoplasmic reticulum membrane in eukaryotes or the plasma membrane of bacteria) via the translocon complex, and the translocon inserts the TM helices into the membrane in sequence order (usually; Sadlish et al. 2005; White and von Heijne 2008). Experimental evidence suggests that the translocon can measure the hydrophobicities of helices to determine which are hydrophobic enough to enter the membrane rather than be secreted, and the associated natural translocon hydrophobicity scale has been derived (Hessa et al. 2005, 2007). Interestingly, according to this scale, a significant fraction of TM helices are insufficiently hydrophobic to be inserted by the translocon as isolated helices but must be inserted in sequence context with their immediately neighboring loops and helices (Hedin et al. 2010). On occasion this is also insufficient, and a more complex interplay between the protein and the translocon must be at work (Hedin et al. 2010). For the folding of some proteins, there is evidence for a much more elaborate and active role for the translocon–ribosome complex than is postulated in passive sequential-insertion models (Kida et al. 2007; Skach 2007; Pitonzo et al. 2009). Clearly, understanding the full function and mechanisms of the translocon will be an important challenge for years to come. Our chief interest here will be the influence of the translocon on the likely contact order of the TM helices and some of the possible structural consequences for the full helix bundles.

2 Overview of non-interhelical stabilizing forces and natural constraints

We first sketch some important non-interhelical interactions. These constraints and interactions will act cooperatively with the interhelical ones that will be our focus.

2.1 Membrane constraints and interactions

The membrane creates some of the foremost natural constraints for TM helical proteins. First, the membrane greatly limits the amino acid composition of a TM helical protein: the protein must be sufficiently hydrophobic to insert into the membrane, yet it must also accommodate the membrane's polar headgroup region. Second, we have the following known geometric constraints.

Beyond these types of protein–membrane constraints, some TM helical proteins have evolved to respond to subtle changes in the membrane including lateral pressure, curvature, lipid composition, and phase, etc. (Perozo et al. 2002; Lundbaek et al. 2010).

2.1.1 Hydrophobic mismatch

A TM helical protein must avoid hydrophobic mismatch with the membrane: the hydrophobic stretch of each TM helix will usually position itself to match the hydrophobic thickness of the membrane. Thus once in the membrane, a helix will usually have a restricted tilt angle with respect to the membrane normal. The specifics are being studied both experimentally and computationally (Bond et al. 2007; Krishnakumar and London 2007; Holt and Killian 2010).

2.1.2 Specific flanking and anchoring interactions with polar headgroups

Residues with a mixed polar/apolar character are common in interfacial regions, and both basic and aromatic residues are reported to form specific favorable interactions with lipid headgroups (Ren et al. 1999; Killian and von Heijne 2000; Strandberg et al. 2002; Chamberlain et al. 2004). Trp anchoring is perhaps best established and has been studied experimentally in synthetic peptides in bilayers, where it is reported to inhibit helix tilting (de Planque et al. 2003; Chiang et al. 2005). Better characterization of protein–membrane interaction motifs would be an important advance.

2.1.3 Positive-inside rule

The positive-inside rule states that the cytoplasmic loops of TM helical proteins tend to be enriched in positively charged residues, and so this charge distribution determines the orientation of most TM helical proteins within the membrane (von Heijne 1989). The physical basis for this tendency is not well-understood and is currently being studied (van Klompenburg et al. 1997; Bogdanov et al. 2008).

2.2 Loop constraints

Loops are clearly important to translocon-aided folding and known to be essential in some cases (White and von Heijne 2008; Hedin et al. 2010).

The effect of loops on stability and folding has been studied in depth experimentally for bacteriorhodopsin (bR) and rhodopsin. In the bR studies, loops were systematically clipped and the resulting fragments were observed to reconstitute a native-like structure, but with reduced stability (Huang et al. 1981; Liao et al. 1984; Popot et al. 1987; Marti 1998). Similar studies were conducted for rhodopsin that showed most of its loops could not be clipped without disrupting the folding of the protein (Albert and Litman 1978; Litman 1979; Ridge et al. 1995; Landin et al. 2001).

Many loops are short and stretched, and so impose a significant geometric constraint on the connected helices (Enosh et al. 2004; Tastan et al. 2009).

3 Interhelical interactions and constraints

We can now begin to discuss how iterative folding and amino acid sequence come together to specify stable structures via well-known specific interhelical interactions. The interhelical constraints act cooperatively with the ones discussed above.

3.1 Helix–helix packing

Crick first described how a simplified view of the surface features of an alpha-helix could be seen to restrict the crossing angle of a close-packed pair of helices (Crick 1953). If the side chains are considered as simple knobs and the spaces between the knobs as holes, then if a pair of helices is to be brought into extended close contact, a series of the knobs must fit into corresponding hole regions. From the geometry of alpha helices and simplifying assumptions, he derived crossing angles for both parallel and antiparallel helices that would enable this type of packing. While that was a groundbreaking analysis, to fully describe helix–helix packing requires a more nuanced approach, as one can see from the more varied helix packing in solved structures.

Walters and DeGrado (2006) have made a very thorough analysis of helix–helix packing in solved TM helical structures. They selected 445 helix pairs from 32 high-resolution protein structures, aligned each possible pair, and clustered them so that each member of a cluster was within 1.5 Å Cα rmsd of a reference centroid pair (the rmsds were computed over 10–14 residue stretches of the TM helices). They found that 29% of pairs fell into the most populous cluster, 74% of pairs fell into the top five most populous clusters, and 90% fell within the top 14. They also found significant amino acid propensities for specific positions in some helix pair clusters. We will reexamine the clusters from a slightly different perspective in a later section.

3.2 Motifs and stabilizing specific interactions

Statistical studies of sequences, examination of solved structures, theoretical analysis, and various types of experiments have led to the description of a small number of common significantly stabilizing interhelical interaction types. It has become common practice for scientists to spotlight where these kinds of interactions occur when introducing new experimental structures or models of TM helical proteins. These are:

3.2.1 Packing motifs

These are usually composed of close knob-in-hole type packed residues each with a side chain of limited conformational flexibility. Thus each such residue can fill a cavity and make van der Waals (VDW) and sometimes polar contacts without significant entropic losses. The most common of these include the famous GxxxG, leucine zippers, and variants of both (Gurezka et al. 1999; Russ and Engelman 2000; Senes et al. 2001; Schneider and Engelman 2004). There are other less common ones, e.g., using proline packing (Senes et al. 2004).

3.2.2 Hydrogen bonds

Although interhelical hydrogen bonds appear to be weaker on average than was once thought, their strength varies greatly according to environment, and they can be significantly stabilizing (Zhou et al. 2000; Gratkowski et al. 2001; Arbely and Arkin 2004; Joh et al. 2008). Residues participating in such hydrogen bonds are often highly conserved (Hildebrand et al. 2008).

3.2.3 Aromatic interactions

The edges of aromatic rings can be considered as weak donors and acceptors. Resulting aromatic interactions include the well-established and important cation–pi interactions, aromatic stacking, edge-to-face aromatic–aromatic interactions, and interactions with polar atoms where the edge acts as donor and the center as acceptor (Dougherty 1996; Johnson et al. 2007; Sal-Man et al. 2007; Nanda and Schmiedekamp 2008).

3.2.4 Salt bridges

Rare in TM helical proteins, but can be significantly stabilizing (Honig and Hubbell 1984).

3.3 The five types of specific stabilizing interhelical interactions considered

We will focus on five types of stabilizing interhelical interactions when analyzing structures. All of the interactions lie within or very close to the inferred hydrocarbon region. Three are polar: hydrogen bonds, salt bridges, and some aromatic interactions. Two are packing interactions: small residue (G/A/S/C) as knob in close knob-in-hole packing and I/V/L/T as knob in close knob-in-hole packing in I/V/L/T contact patches. For I/V/L/T patch packing, at least one of the surrounding hole residues must be I/V/L/T, along with some other restrictions on the hole residue types (Harrington and Ben-Tal 2009). Note that these packing interactions have been defined on a residue basis, so common packing motifs would consist of multiple such interactions.

For each interaction type, there is a fixed interaction geometry. These interaction geometries are fixed sets of geometric conditions that must be satisfied if the donor and acceptor are interacting (e.g., the usual conditions for a hydrogen bond; for the two packing interactions, the knob is considered the donor and the hole the acceptor; Harrington and Ben-Tal 2009).

3.4 Structural hot spots

An idea related to the above list of interactions is that of what is sometimes called a structural hot spot: a residue or residues making particularly favorable contributions to stability (Bogan and Thorn 1998; Fleming and Engelman 2001). We will call a residue "structurally hot" or "particularly stabilizing" when it makes interhelical contacts so that its contribution to the stability of the conformation is especially favorable, and much more favorable than would be expected for a typical residue in a typical conformation of the protein. It has been suggested that residues likely to be structurally hot make numerous favorable contacts: some VDW, but also other more specific favorable interactions (Gao and Li 2009). Such residues are sometimes termed "hub residues" (Pabuwal and Li 2008, 2009); e.g., a polar residue that makes numerous VDW contacts while also hydrogen bonding. This characterization of structurally hot residues has experimental support, as will be described in the next section, and indicates why some of the weaker specific interactions included in our list can still play a critical role in creating structural hot spots. This is clear if the weak interactions are likely stronger than generic VDW contacts would be for the atoms in the interactions.

The structural hot spot concept relates to the motif-type analysis of structures: one can try to describe a set of geometric conditions for a set of residues to satisfy (often specifying residue types of interacting partners) so that the resulting contribution of the residues in the motif is very likely significantly favorable to the stability of the structure. For the common motifs, it will usually be true that at least one residue of the motif will contribute significantly more to stability than is expected for a typical residue. To facilitate structural analysis, it is convenient to take the various packing motifs apart into the packing of their residues.

Theoretically, one would like a list of simple structural conditions to place on amino acids so that the probability that they make a significantly favorable contribution to stability is much higher than that of a typical residue making typical contacts. In practice, consideration of the types of favorable interactions described above and characterizations of common motifs are extremely useful approximations to an ideal set of descriptors. They are particularly useful because they do not depend on complex tertiary contacts. Two things should be kept in mind when using these simplified conditions. First, the energies of the interactions can vary greatly, and in the case of weak hydrogen bonds or other polar interactions, the bond by itself does not make a participating residue "structurally hot". But the existence of even a weak bond increases the probability that a participating residue makes a significant contribution to the stability of the conformation. Second, even optimally defined conditions would have a probabilistic implication of stability, and a residue satisfying them might not be stabilizing.

In reality, not all structurally hot residues are involved in interactions of one of the five types. Conspicuously absent are residues that simply make many VDW contacts. These will primarily be residues with many tertiary contacts, and such structural hot spots could be recognized only when the protein is assembled to a sufficient extent. This limits their use during most steps of iterative translocon-aided folding or structure prediction assembly.

3.5 Experimental data on residue contributions to stabilization

Faham et al. (2004) investigated the contributions of residues to stability in bR by systematically mutating the residues of helix B to alanine and measuring the thermodynamic stability of the mutants using an unfolding assay. We will interpret their data in terms of the five types of stabilizing interactions. (Faham et al. did not present their results in these terms but instead emphasized VDW interactions in their interpretation of the data.)

Each mutant was classified as either severely destabilized, moderately destabilized, minimally altered, or stabilized. Of the 24 residues mutated, 17 could be said to have some kind of interhelical contact (of greatly varying extents) with one or more of the other bR helices, and we will restrict our attention to those. (The structure used for our analysis is 1C3W.) The notation used below gives the interactions in donor–acceptor form.

Of the 17, four were severely destabilizing: F42A, I45A, T46A, and Y57A. All of these residues make particularly stabilizing interhelical interactions of the five types, and two of them make at least two, in line with the "hub residue" type of structural hot spot. The mutations do not create favorable interactions of the five types, but do destroy the native ones. (F42: 42-CD1-96-OD1, 42-CE1-96-O; I45 I/V/L/T patch packing; T46:46-OG1-96-OD2; Y57: 57-CD2-13-O, 57-OH-212-OD2, 57-CE1-212-OD2.)

Five of the 17 mutants were moderately destabilizing: Y43A, T47A, I52A, F54A, and M60A. The first four of these also make at least one particularly stabilizing inter-helical interaction of the five types. Again, the mutations do not create favorable in-teractions. (Y43:30-NZ-43-OH, 43-CD1-27-O (borderline); T47: I/V/L/T patch packing, 47-OG1-27-O (water bridged); I52: I/V/L/T patch packing (borderline); F54: 54-CD1-17-O.)

The exception is M60, which by its residue type cannot participate in an interac-tion of one of the five types. Interestingly, it makes extensive VDW contact with Y57 in the native structure. Y57A is the most destabilizing mutation of all, and Y57 does make three particularly stabilizing interactions as was described. It appears that the side chain of M60 guides Y57 to make those interactions, thereby further stabilizing the native position of Y57.

The five minimally altered mutants of the 17 were: L48A, P50A, T55A, L58A, and S59A. With the exception of P50, these residues make few contacts of any kind. While we did not include proline packing in our list of favorable interactions, it would have been fair to do so. Since proline close packing would be considered favorable because it has a small side chain of restricted mobility, one would not expect P50A to much change stability; it is a moderate substitution. None of the other mutations would either add or remove a favorable interaction of the five types.

The three stabilized mutants of the 17 were: V49A, M56A, and L61A. Both V49 and L61 act as close-packed knobs in favorable I/V/L/T patch interactions. The substitution changes both interactions to a small residue knob-in-hole interaction, and so V49A and L61A amount to a substitution of one kind of favorable packing interaction for another. That alanine packing is superior is consistent with results on zipper motifs in TM helix pairs suggesting that small residue close knob-in-hole packing is more stabilizing than I/V patch packing (Zhang et al. 2009). M56 does not participate in any of the interactions of the five types, but M56A will convert it to a small residue in close knob-in-hole packing. Thus this substitution adds a par-ticularly favorable packing interaction.

Overall, this data and above analysis support the stabilizing significance of the interactions of the five types.

3.6 Particularly stabilizing interactions as geometric constraints

By examining solved structures for these kinds of particularly favorable interhelical interactions, one can see that they seem to be distributed in a surprisingly meaningful way: the interactions appear to tightly constrain the packing of the helices in most solved TM helical proteins. To put this another way, it appears that if we significantly perturb the helix positions, we necessarily break some interactions in the set. If true, one might say a complete set of determining constraints/interactions of the five types has evolved to nearly fix each native state backbone conformation. We call sets of the five types of interhelical interactions "determining sets" when they nearly

Cluster 1

Cluster 2

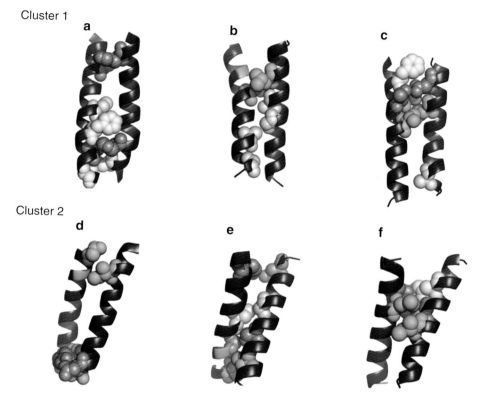

Fig. 1. **Unrelated members of the most common TM helix pair folds stabilized by diverse determining sets.** The five types of interhelical interactions are displayed as follows. (Only four types occur in these structures: there are no salt bridges.) Residues in a hydrogen bond are colored orange; if a residue's side chain (rather than a backbone atom) is in a hydrogen bond its atoms are shown as orange spheres. Residues in aromatic interactions are colored yellow; if their side chains participate in the interaction, the side chain atoms are shown as spheres. The knob atoms of G/A/S/C small close knob-in-hole packing are shown as wheat colored spheres. The knob atoms of I/V/L/T close knob-in-hole packing in I/V/L/T patches are shown as cyan (bright blue) spheres. The corresponding hole residues are shown as spheres if they are I/V/L/T in close contact with the knob residue; otherwise if the hole residue has restricted side chain conformations it is also shown in cyan. We can see unrelated helix pairs from the same fold with diverse determining sets. Upper panels: pairs from the top Walters–DeGrado cluster (**a**) 1Q90: cytochrome B6-F. (**b**) 1H2S: sensory rhodopsin II. (**c**) 1C3W: bR (a different helix pair than the homologous one for 1H2S). Lower panels: pairs from the next most populous cluster (**d**) 1OCR: cytochrome C oxidase. (**e**) 1RH5: translocase SecE subunit. (**f**) 1OKC: ADP/ATP carrier protein.

fix geometrically the packing of the helix backbones. (Our terminology is based on the geometric meaning of "determine": to specify position, to fix.)

We will discuss how these observations relate to some fundamental questions of protein folding and describe our work to establish them. But we first look at some simple examples.

3.7 Helix pairs revisited

The Walters–DeGrado clusters provide examples of diverse ways in which the five types of interactions can be distributed to fix the same helix-pair fold. The helix pairs within a cluster usually come from unrelated proteins or from different parts of the same protein. As members of the same cluster, they must have similar conformations for the aligned 10–14 residue region. Since most TM helices are at least 20 residues long, the position of this region can vary, and so for some clusters not all the full helix pairs look similar.

We have chosen pairs that do look similar and stable to show the diverse ways that a fold can be stabilized. In Fig. 1, we see three members from each of the two most populous clusters with their determining sets displayed. Imagine perturbing the positions of the helices to see how little one can move the helices without disrupting these interactions.

3.8 Constraint perspective and underlying rigid-body geometry

Both helix packing and the constraining effect of the five types of interactions can be better understood if some facts about the geometry of rigid bodies are kept in mind.

Any rigid body's position in space can be specified by the positions of any three non-collinear points on the body. If the exact positions of those three points are unknown, but we do know that they must each lie within three given regions in space, then we can obtain an initial ensemble of positions of the body by placing grids on those three regions and systematically selecting these points to give the positions of the three points on the body. These three positions then fix the position of the body itself. If there are additional restrictions on the positions in space of any other points on the body, then we can check the initial ensemble of positions of the body and remove any positions from the ensemble that do not meet those restrictions. By choosing sufficiently fine grids, one can find to any desired accuracy how the specified regions constrain the position of the body.

This approach can be adapted to build the combined piece ensemble of two pieces constrained by a set of any type of interactions with fixed known interaction regions. Three of these interactions and their associated regions can be used to position one piece relative to the other and build the initial combined piece ensemble. If there are

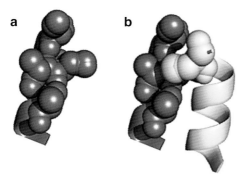

Fig. 2.   **Knob-in-hole.** (**a**) The residues j, $j+3$, $j+4$, and $j+7$ on an alpha helix with reduced representations of their side chains with their atoms shown as spheres. The numbering starts from the top in this picture, and the space surrounded by these residues is called a hole. (**b**) An example of interhelical knob-in-hole packing. The knob is in yellow and the hole in blue.

more interactions of these types or additional geometric conditions (as is the case, e.g., for a hydrogen bond), the initial conformations can be checked and discarded if they do not meet these additional conditions. For our applications, four interactions would tend to constrain the conformations well.

For analysis of packing, we consider knob-in-hole interactions. Each knob is a reduced rigid representative of a side chain based on amino acid type, and each hole is a predefined region (Fig. 2). Each hole is given by the space between a set of helix

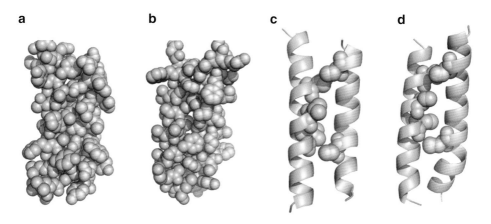

Fig. 3.   **Equivalently packed helices.** These unrelated helix pairs have very similar folds. Fully spacefilled models: (**a**) helix pair from bR, (1C3W); (**b**) helix pair from cytochrome B6-F (1Q90). Note how different the side chains are in panels (**a**, **b**). In panels (**c**, **d**) we see the common packing of knobs-in-holes in the two structures despite their sequence differences. The knobs are fixed reduced representatives of side chains based on residue type. The five knobs shown spacefilled are packed in a nearly equivalent way in (**c**) the helix pair from 1C3W, also shown in (**a**, **d**) the helix pair from 1Q90, also shown in (**b**).

215

residues $i, i + 3, i + 4, i + 7$, but subdivided into three regions (Harrington and Ben-Tal 2009). Four knobs-in-holes would usually constrain a conformation of a helix pair quite well. But the knobs of different amino acids are similar enough so that this is often true even when comparing the packing of helices of different amino acid composition. That is, if four corresponding knobs pack into the same four corresponding holes, the two helix pairs will usually have similar conformations, especially around this region. In Fig. 3, we see unrelated helix pairs with five knobs packed in the same way in both pairs. This can be seen in the helix pair clusters, except there is no condition that knobs pack into corresponding positions on the two pairs of helices, and so the result is local to the packing region.

3.9 Iterative reassembly of full TM helix bundles using interactions of the five types

To analyze the interactions of the five types as constraints, we used rigid motions to iteratively reassemble the helix bundle backbone of each protein using only its set of the five types of interhelical interactions, predefined interaction geometries, and individual helix backbones. Beginning with N rigid separate pieces, initially the individual helices, we fit two together and so obtained a new set of N–1 rigid pieces. After repeating this N–1 times, there is one piece at the end, the assembled structure (actually an ensemble of structures as explained above).

From each solved structure, the backbone conformations of the individual helices and the side chain conformations of those residues with a side chain atom explicitly in an interhelical interaction of one of the five types were taken. Those native side chain conformations are fixed and rigid. The side chains were not taken for residues in the two packing interactions. All other side chains had a fixed, rigid, reduced representation based on residue type that is intended to give the obstruction created by a side chain of that residue type irrespective of rotameric state; they were not derived from the native structures.

The scoring of the structures depended only on overlap penalties and the geometric conditions imposed by the interactions. It does not approximate energy: in particular, a VDW term was not used. For details, see Harrington and Ben-Tal (2009).

The order of reassembly was chosen to mimic a plausible translocon-guided folding pathway for each protein, and so we attempted to rebuild the structures from the N-terminus in a sequence order preserving fashion.

For example, here is our iterative assembly for the voltage sensor, 1ORS. At each step, we put two rigid pieces together using the interhelical interactions between them to produce a new fixed piece (actually an ensemble as described before), as shown in Fig. 4. We first assemble the first two helices, then add 3-a to the single

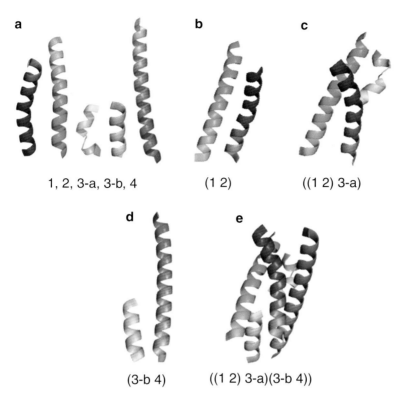

Fig. 4. Assembly order for voltage sensor. An example of our iterative sequence order respecting assembly for the voltage sensor 1ORS. We begin with the individual helices 1, 2, 3-a, 3-b, 4, and assemble iteratively in the order shown. (**a**) The helices to be assembled in sequence order. (**b**) The first two helices assembled, (1 2). (**c**) The third (half) helix is assembled with the first two, ((1 2) 3-a). (**d**) The last two helices assembled together, (3-b 4). (**e**) The piece made up of the first three helices and the piece made up of the last two are assembled together to build the full structure, (((1 2) 3-a) (3-b 4)). Figure adapted from Harrington and Ben-Tal (2009).

piece (1 2). At this point, there are insufficiently many interactions to add 3-b to the first piece ((1 2) 3-a), so we next assemble 3-b and 4, and finally put ((1 2) 3-a) and (3-b 4) together to obtain the full structure (((1 2) 3-a) (3-b 4)).

3.10 The sets of the five types of particularly favorable interactions determine the packing of helices in the native structures of a diverse test set

For a diverse test set of 15 TM helical proteins, the structures rebuilt in the fashion outlined above had an average ensemble-average Cα rmsd from the native of 1.03 Å

(Harrington and Ben-Tal 2009). Furthermore, with the exception of aquaporin, the structures could be rebuilt in a sequence order preserving fashion consistent with translocon-aided folding. In the case of aquaporin, the half-helices needed to be assembled slightly out of order. This might relate to experimental results indicating some unusual insertion behavior of half-helices in general and helices in other aquaporins in particular (Pitonzo and Skach 2006; Jaud et al. 2009).

Determining sets of interactions of the five types seem to be a very common structural feature of solved TM helical proteins, and for good physical reasons as we will explain. But there are proteins without them (e.g., proteins with large prosthetic groups). For a discussion, see Harrington and Ben-Tal (2009).

3.11 Distribution of particularly stabilizing residues, folding funnels, and the construction of low-energy minima

If the residues participating in these types of interactions are likely to make particularly favorable contributions to stability, then their distribution in determining sets of interactions partially explains how sequence specifies structure. These sets of interactions help to create low-energy minima for two reasons. First, the abundance of these particularly favorable interactions would tend to act to make the structure a low-energy one. Second, when the backbone positions of the helices are significantly perturbed, some of the determining set of the interactions will necessarily be broken. At the very least and for very few perturbations, some side chains must be flipped and so rotameric barriers crossed. If we assume the interaction energies are strong enough, it will be difficult to compensate for the lost interactions of the five types given their geometric and partner specificity and the rarity of possible participants. Thus the energies of the perturbed structures would tend to be higher. In contrast, one could not usually say the same of a "determining set of VDW interactions" because of the density and promiscuity of VDW interactions.

The iterative assembly (consistent with translocon-aided folding) and the determining sets of interactions can also be seen as a geometric recipe for creating folding funnels. The interactions of the five types are supposed to be individually and locally superior to generic contacts and so can successively funnel and collectively trap the native backbone. That this could be done in a controlled iterative way aided by the translocon makes the process much simpler. The native backbone conformations (and sub-conformations) usually have many interactions of the five types that can stabilize them in addition to the ones that appear in the solved structures since side chains that participate in these interactions can adopt different conformations and form different interactions. These additional interactions could further aid the funneling process.

3.12 Cooperativity with packing

Conformations of proteins are restricted by simple packing rules of the type we have seen for helix pairs. Between every pair of pieces (helices or subbundles) we used for reassembly, there must be at least four interactions of the five types. But if we instead consider either knob-in-hole or aromatic interactions (for loosely packed proteins) between those same pieces, we find that there are also at least three of these types of packing interactions (overwhelmingly knob-in-hole; Harrington 2009, unpublished data). Thus these native conformations are also quite constrained by simple packing interactions (although not as constrained as by the interactions of the five types).

3.13 Static structures versus ensembles

The sets of interactions of the five types were derived from crystal structures. In reality, an ensemble of structures underlies any crystal structure, which complicates the analysis. Due both to the limits of resolution and the underlying multiplicity of structures, it can be difficult to read the set of these interactions from a structure. To deal with this, we added error terms to the geometric conditions; to fully address it would require native state ensembles for all of the crystal structures. But the fact that for a native backbone conformation there can be multiple determining sets due to, e.g., bond switching (residues in hydrogen bonds changing partners or rotamers) does not contradict our analysis. If a determining set of interactions is fixed, the backbone conformations are highly constrained but can slightly jitter without breaking the interactions. If one began with a different determining set for the same backbone conformation, the two backbone ensembles constrained by the two different determining sets would not be identical, but very similar.

4 Conservation and diversity of determining sets of stabilizing interactions

If such structural importance is attributed to the determining sets of interactions of these types, then what happens in proteins of the same fold? For homologs, do the residues participating in these kinds of interactions have to be conserved or is some diversity possible? What about for remote homologs or proteins with related folds? How many ways are there to fix similar backbone conformations using these types of interactions?

As we have seen for helix pairs, great diversity in the determining sets of interactions is possible for very similar backbone conformations from unrelated proteins (Fig. 1). In homologs, we can also see this diversity.

For homologs, the conservation pattern of residues participating in the determining sets of interactions is mixed. In general, buried residues are more conserved than exposed ones (Baldwin et al. 1997; Fleishman et al. 2004; Liao et al. 2005;

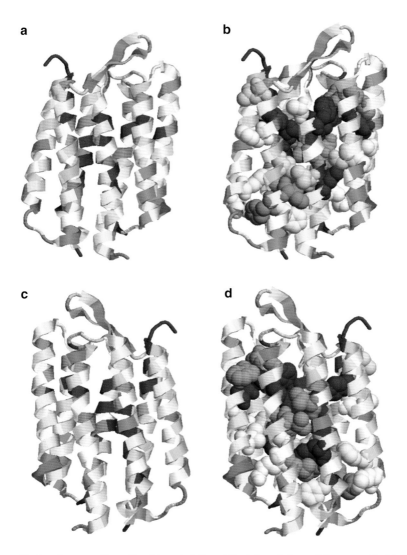

Fig. 5. Conservation of residues in bR and its determining set. The conservation is indicated by color from turquoise to maroon: turquoise means highly variable, and maroon means highly conserved. White is intermediate. (**a**) One view of bR. (**b**) Here the residues involved in the five types of interactions are spacefilled. (**c**) Another view of bR. (**d**) Another view of the residues in the determining set as spacefilled. Thus almost all but the most exposed residues in the determining set of interactions are conserved. The exposed ones can vary.

Park and Helms 2007; Hildebrand et al. 2008). For helical membrane proteins, it has been found that residues in hydrogen bonds are 40% more conserved than exposed residues while buried residues are 25% more conserved than exposed residues (Hildebrand et al. 2008). For many closely packed small residues, substitution by a bulkier residue would require a substantial change in the backbone conformation, and so such residues will tend to be conserved in the same fold for that reason alone. Buried residues in the determining sets tend to be very highly conserved. For more exposed interactions in the determining sets, diverse alternative sets of these constraints are commonly seen, and such residues less conserved (Liu et al. 2004). We will now look at this phenomenon in bR and its homologs.

4.1 Conservation and diversity of the determining sets of interactions of bR

We will consider the two most conserved categories of residues as classified by ConSurf: highly conserved and conserved (Fig. 5; Glaser et al. 2003). In the helical part of bR, there are 17 residues within the inferred hydrocarbon region classified as highly conserved (the top category), and 11 of these participate in the determining set of interactions. Of the six remaining, two are helix-kinking prolines, one is K216, which is critical for binding retinal, and one is M60, which we have discussed as a guide for Y57. There are 18 helix residues classified as conserved (the next-to-top category), 11 of which are part of the determining set of interactions. Again, many of those conserved residues not participating in the determining set of interactions are known to be functionally important or related to important secondary structure features.

The less buried residues in the determining sets are much less conserved. In the crystal structures of the homologs of bR, we can see the diversity of the less buried parts of the determining sets of interactions (Fig. 6). Intuitively, it is unsurprising that mutations of more exposed residues are more likely to result in diverse determining sets of interactions simply because there would not be the additional geometric constraints imposed by surrounding residues, so the evolutionary process would not require as many concerted mutations.

5 Determining sets, multiple states, and motion

Determining sets of interactions highly constrain the positions of helix bundle backbones, so how do they relate to the dynamic properties of TM helical proteins? The simplest answer according to the constraint philosophy is that each conformational state (or highly constrained ensemble) corresponds to a different determining set,

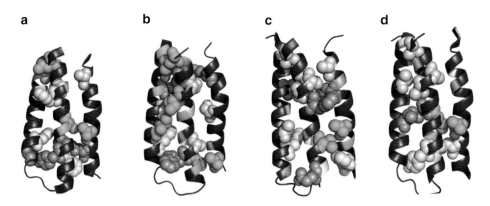

Fig. 6. Comparison of homologs: The same fold with diverse determining sets of interactions.
The residues in the determining sets are spacefilled and color-coded as in Fig. 1. The panels (**a**, **b**) show bacteriorhodopsin and halorhodopsin. (**a**) Bacteriorhodopsin (1C3W). (**b**) Halorhodopsin (1E12). The panels (**c**, **d**) show sensory rhodopsins II. (**c**) Sensory rhodopsin II, Anabaena (1XIO). (**d**) Sensory rhodopsin II, N. pharaonis (1H68). Figure adapted from Harrington and Ben-Tal (2009).

and motion would result from a collection of these states. There should also be a reasonable transition between states, which from the constraint perspective would mean a way to switch smoothly between the different determining sets of interactions corresponding to the two states. Suggested mechanisms for these transitions include deformation of hydrogen bonds, with non-native bonds along the transitions or rotameric flips, and sliding in interfaces dominated by small residues (Perozo et al. 2002; Curran and Engelman 2003; Hildebrand et al. 2008; Gardino et al. 2009).

5.1 Multiple states and motion in the ErbB family

The epidermal growth factor family of receptor tyrosine kinases is an interesting case in point. The members of this family (ErbB1, ErbB2, ErbB3, and ErbB4) play critical roles in a variety of physiological processes and their malfunction has been associated with many cancers. Each member has the same overall components: an extracellular ligand-binding domain connected to a single TM helix, which is connected to a cytoplasmic tyrosine kinase domain. Usually the formation of hetero- and homodimers of the TM helices in this family is induced by the binding of ligands to the extracellular domain. A ligand specific to ErbB2 has not been found, but it can be affected by the ligand-binding of the other family members. Furthermore, it does appear to form active homodimers on its own. Dimerization can trigger the tyrosine kinase activity of the cytoplasmic domain, and so it has a crucial functional role for these receptors.

```
ErbB1    I A T G M V G A L L L L L V V A L G I G L F M
ErbB2    S I V S A V V G I L L V V V L G V V F G I L I
ErbB3    M A L T V I A G L V V I F M M L G G T F L
ErbB4    I A A G V I G G L F I L V I V G L T F A V Y V
```

Fig. 7. Multiple sequence alignment around the TM segments of the four human ErbB paralogs. The N-terminal GxxxG-like motif is indicated in yellow and the C-terminal GxxxG-like motif in blue. Note that ErbB3 does not have a C-terminal motif of this type.

The TM dimers of this family make attractive candidates for study because of their simplicity and medical significance. The TM helices all contain similar dimerization motifs near the N-terminus of their helices similar to GxxxG, but actually defined earlier (Sternberg and Gullick 1990). Seven residues later in their helix sequences, for all but ErbB3, there appears to be another dimerization motif related to the famous GxxxG (Fig. 7). This suggests that there are at least two main conformational states in the homo and heterodimers: one corresponding to the N-terminal motif and the other to the C-terminal motif. Using a scoring function based on rewarding the packing of small residues and a grid search for the helix pair conformations, Fleishman et al. (2002) produced model structures for these two states for ErbB2 and proposed a molecular switch for activation based on them. In this switch model, the conformation induced by the N-terminal motif is the active form of ErbB2, and the conformation induced by the C-terminal motif the inactive form. This model was found to explain some known disease-causing mutations in detail, and later the pathway between them was validated by motion-planning methods (Fleishman et al. 2002; Enosh et al. 2007). Both these proposed conformations, as well as the proposed pathway between them, are consistent with the determining set perspective. The two GxxxG-like motifs as well as the nearby polar (N-terminal), aromatic (C-terminal), and V/I/L/T residues can all participate in the five types of interactions and highly constrain the conformations. The proposed intermediate states feature extensive close knob-in-hole V/I/L/T patch packing and some small residue packing, and so are also in line with the determining set philosophy of motion.

These ideas have experimental support. Solution NMR structures for the TM dimer of ErbB2 have been found, and they closely agree with the earlier proposed model for the active state (Bocharov et al. 2007). The conformations use the N-terminal dimerization motif as expected, and there are interhelical hydrogen bonds between the hydroxyl groups of Ser656 and between the hydroxyl groups of Thr652. These interhelical hydrogen bonds are transient and vary among the conformations in the ensemble (Fig. 8). Additionally, an aromatic–aromatic edge-to-face interhelical in-

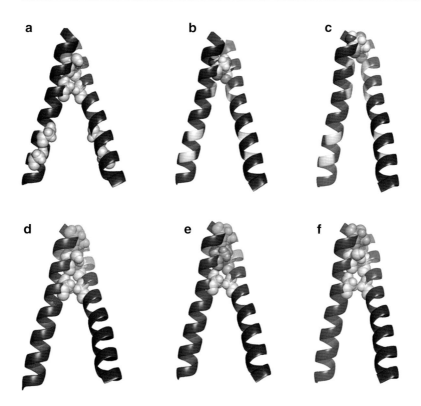

Fig. 8. **NMR structures of the active state of ErbB2.** The stabilizing interactions for the ensemble of the active state of the ErbB2 dimer (2JWA). For both upper and lower panels, three different models from the ensemble are shown. Upper panel: (**a**) shows the two GxxxG-like dimerization motifs spacefilled and colored in wheat. The unused C-terminal motif is believed to mediate dimerization of the inactive state. (**b**) Transient hydrogen bonds between the hydroxyl groups of S656 are seen; here is one displayed spacefilled. (**c**) Transient hydrogen bonds between the hydroxyl groups of T652 are seen; here one is shown spacefilled. Lower panel: Since the hydrogen bonds vary among the ensemble members, their determining sets differ. However, both S656 and T652 are in particularly stabilizing knob-in-hole packing interactions whether or not they hydrogen bond, so the conformations are well-constrained even without the hydrogen bonds. The determining sets of interactions are color-coded as described for Fig. 1. (**d**) Neither S656 nor T652 form hydrogen bonds in this conformation, so there is only small residue packing and I//V/L/T packing. (**e**) There is a S656–S656 hydrogen bond shown in orange in this conformation. (**f**) There is a T652–T652 bond shown in orange in this conformation.

teraction is reported. All of this is consistent with the determining set perspective we have described.

There is also experimental support for the proposed inactive state of ErbB2. Many experiments have tested the suspected dimerization motifs in the ErbB family. Escher et al. (2009) studied the dimerization capabilities in a biological membrane of both N- and C-terminal dimerization motifs for both homo- and heterodimers.

The N- and C-terminal motifs were studied separately by dividing each TM helix into two parts to create two new TM segments. One of the new TM segments contained the N-terminal motif only and the other contained the C-terminal motif only, such that the two motifs occupied equivalent positions on their respective segments. For ErbB2, it was found that these C-terminal motif segments dimerized, supporting this putative dimerization motif and hence also the putative inactive state structure.

There is very active study of the entire ErbB family; for reviews, see (Landau and Ben-Tal 2008; Hynes and MacDonald 2009; Lemmon 2009). One can hope that analysis of the type done successfully for ErbB2 will also work for the other possible states of the homo- and heterodimers of this family. Perhaps such methods can also be applied to other tyrosine kinase receptors.

6 Conclusion

From the earliest days of structural biology, the notion of energetically important interactions as geometric constraints has been key to many classic discoveries. We have argued that the idea also sheds light on the complex problems of understanding TM helical proteins today. The modern emphasis on multiple states and ensembles of structures as the key to function (Henzler-Wildman and Kern 2007) has a direct connection to old-fashioned model building. Just as model-building was guided by physically important interactions as constraints, these multiple states and the transitions between them can be seen in part as geometrically created and mediated by determining sets of well-known favorable interactions. The folding and dynamics of TM helical proteins, their structure prediction and eventual design can all be seen more clearly in this light, despite our imperfect knowledge of these stabilizing interactions, especially their energetics. We believe that these ideas form the foundation for new top–down algorithms for structure and motion prediction which could bridge the gap to more detailed bottom–up approaches such as molecular dynamics simulations.

References

Albert AD and Litman BJ (1978) Independent structural domains in the membrane protein bovine rhodopsin. Biochemistry 17: 3893–3900

Arbely E and Arkin I (2004) Experimental measurement of the strength of a Cα–H–O bond in a lipid bilayer. J Am Chem Soc 126: 5362-5363

Baldwin JM, Schertler GF, Unger VM (1997) An α-carbon template for the transmembrane helices in the rhodopsin family of G-protein-coupled receptors. J Mol Biol 272: 144–164

Bocharov EV, Mineev KS, Volynsky PE, et al. (2007) Spatial structure of the dimeric transmembrane domain of the growth factor receptor ErbB2 presumably corresponding to the receptor active state. J Biol Chem 283: 6950–6956

Bogan AA and Thorn KS (1998) Anatomy of hot spots in protein interfaces. J Mol Biol 280: 1–9

Bogdanov M, Xie J, Heacock P, et al. (2008) To flip or not to flip: lipid–protein charge interactions are a determinant of final membrane topology. J Cell Biol 182: 925–935

Bond PJ, Holyoake J, Ivetac A, et al. (2007) Coarse-grained molecular dynamics simulations of membrane proteins and peptides. J Struct Biol 157: 593–605

Bowie JU (2005) Solving the membrane protein folding problem. Nature 438: 581–589

Chamberlain AK, Lee Y, Kim S, et al. (2004) Snorkeling preferences foster an amino acid composition bias in transmembrane helices. J Mol Biol 339: 471–479

Chiang CS, Shirinian L, Sukharev S (2005) Capping transmembrane helices of MscL with aromatic residues changes channel response to membrane stretch. Biochemistry 44: 12589–12597

Curran AR and Engelman DM (2003) Sequence motifs, polar interactions and conformational changes in helical membrane proteins. Curr Opin Struct Biol 13: 412–417

Crick F (1953) The packing of α-helices: simple coiled-coils. Acta Crysta 6: 689–697

de Planque MRR, Bonev BB, Demmers JAA, et al. (2003) Interfacial anchor properties of tryptophan residues in transmembrane peptides can dominate over hydrophobic matching effects in peptide-lipid interactions. Biochemistry 42: 5341–5348

Dougherty D (1996) Cation-pi interactions in chemistry and biology: a new view Benzene, Phe, Tyr, and Trp. Science 271: 163–168

Enosh A, Fleishman SJ, Ben-Tal N, et al. (2004) Assigning transmembrane segments to helices in inter-mediate-resolution structures. Bioinformatics 20: i122–i129

Enosh A, Fleishman SJ, Ben-Tal N, et al. (2007) Prediction and simulation of motion in pairs of trans-membrane alpha-helices. Bioinformatics 23(2): e212–e218

Escher C, Cymer F, Schneider D (2009) Two GxxxG-like motifs facilitate promiscuous interactions of the human ErbB transmembrane domains. J Mol Biol 389: 10–16

Faham S, Yang D, Bare E, et al. (2004) Side-chain contributions to membrane protein structure and stability. J Mol Biol 335: 297–305

Fleming KG and Engelman DM (2001) Specificity in transmembrane helix–helix interactions can de-fine a hierarchy of stability of sequence variants. Proc Natl Acad Sci USA 98: 14340–14344.

Fleishman SJ, Schlessinger J, Ben-Tal N (2002) A putative molecular-activation switch in the trans-membrane domain of erbB2. Proc Natl Acad Sci USA 99: 15937–15940

Fleishman SJ, Harrington S, Friesner RA, et al. (2004) An automatic method for predicting transmem-brane protein structures using cryo-EM and evolutionary data. Biophys J 87: 3448–3459

Gao J and Li Z (2009) Comparing four different approaches for the determination of inter-residue in-teractions provides insight for the structure prediction of helical membrane proteins. Biopolymers 91: 547–556

Gardino A, Villali J, Kivenson A, et al. (2009) Transient non-native hydrogen bonds promote activa-tion of a signaling protein. Cell 139: 1109–1118

Glaser F, Pupko T, Paz I, et al. (2003) ConSurf: identification of functional regions in proteins by sur-face-mapping of phylogenetic information. Bioinformatics 19: 163–164

Gratkowski H, Lear JD, DeGrado WF (2001) Polar side chains drive the association of model trans-membrane peptides. Proc Natl Acad Sci USA 98: 880–885

Gurezka R, Laage R, Brosig B, Langosch DA (1999) Heptad motif of leucine residues found in mem-brane proteins can drive self-assembly of artificial transmembrane segments. J Biol Chem 274: 9265–9270

Harrington SE and Ben-Tal N (2009) Structural determinants of transmembrane helical proteins. Structure 17: 1092–1103

Hedin L, Ojemalm K, Bernsel A, et al. (2010) Membrane insertion of marginally hydrophobic transmembrane helices depends on sequence context. J Mol Biol 396: 221–229

Henzler-Wildman K and Kern D (2007) Dynamic personalities of proteins. Nature 450: 964–972

Hessa T, Kim H, Bihlmaier K, et al. (2005) Recognition of transmembrane helices by the endoplasmic reticulum translocon. Nature 433: 377–381

Hessa T, Meindl-Beinker NM, Bernsel A, et al. (2007) Molecular code for transmembrane-helix recognition by the Sec61 translocon. Nature 450: 1026–1030

Hildebrand PW, Gunther S, Goede A, et al. (2008) Hydrogen-bonding and packing features of membrane proteins: functional implications. Biophys J 94: 1945–1953

Holt A and Killian JA (2009) Orientation and dynamics of transmembrane peptides: the power of simple models. Eur Biophys J 39: 609–621

Honig BH and Hubbell WL (1984) Stability of "salt bridges" in membrane proteins. Proc Natl Acad Sci USA 81: 5412–5416

Huang KS, Bayley H, Liao MJ, et al. (1981) Refolding of an integral membrane protein. Denaturation, renaturation, and reconstitution of intact bacteriorhodopsin and two proteo-lytic fragments. J Biol Chem 256: 3802–3809

Hynes NE and MacDonald G (2009) ErbB receptors and signaling pathways in cancer. Curr Opin Cell Biol 21: 177–184

Jaud S, Fernandez-Vidal M, Nilsson I, et al. (2009) Insertion of short transmembrane helices by the Sec61 translocon. Proc Natl Acad Sci USA 106: 11588–11593

Joh NH, Min A, Faham S, et al. (2008) Modest stabilization by most hydrogen-bonded side-chain interactions in membrane proteins. Nature 453: 1266–1270

Johnson RM, Hecht K, Deber CM (2007) Aromatic and cation–pi interactions enhance helix–helix association in a membrane environment. Biochemistry 46: 9208–9214

Kida Y, Morimoto F, Sakaguchi M (2007) Two translocating hydrophilic segments of a nascent chain span the ER membrane during multispanning protein topogenesis. J Cell Biol 179: 1441–1452

Killian JA and von Heijne G (2000) How proteins adapt to a membrane-water interface. Trends Biochem Sci 25: 429–434

Krishnakumar SS and London E (2007) Effect of sequence hydrophobicity and bilayer width upon the minimum length required for he formation of transmembrane helices in membranes. J Mol Biol 374: 671–687

Landau M and Ben-Tal N (2008) Dynamic equilibrium between multiple active and inactive conformations explains regulation and oncogenic mutations in ErbB receptors. Biochim Biophys Acta 1785: 12–31

Landin JS, Katragadda M, Albert AD (2001) Thermal destabilization of rhodopsin and opsin by proteolytic cleavage in bovine rod outer segment disk membranes. Biochemistry 40: 11176–11183

Lemmon MA (2009) Ligand-induced ErbB receptor dimerization. Exp Cell Res 15: 638–648

Liao H, Yeh W, Chiang D, et al. (2005) Protein sequence entropy is closely related to packing density and hydrophobicity. Protein Eng Des Sel 18: 59–64

Liao MJ, Huang KS, Khorana HG (1984) Regeneration of native bacteriorhodopsin structure from fragments. J Biol Chem 259: 4200–4204

Litman BJ (1979) Rhodopsin: its molecular substructure and phospholipid interactions. Photochem Photobiol 29: 671–677

Liu W, Eilers M, Patel AB, et al. (2004) Helix packing moments reveal diversity and conservation in membrane proteins. J Mol Biol 337: 713–729

227

Lundbaek JA, Collingwood SA, Ingolfsson HI, et al. (2010) Lipid bilayer regulation of membrane protein function: gramicidin channels as molecular force probes. J R Soc Interface 7: 373–395

Marti T (1998) Refolding of bacteriorhodopsin from expressed polypeptide fragments. J Biol Chem 273: 9312–9322

Nanda V and Schmiedekamp A (2008) Are aromatic carbon donor hydrogen bonds linear in proteins? Proteins 70: 489–497

Pabuwal V and Li Z (2008) Network pattern of residue packing in helical membrane proteins and its application in membrane protein structure prediction. Prot Eng Des Sel 21: 55–64

Pabuwal V and Li Z (2009) Comparative analysis of the packing topology of structurally important residues in helical membrane and soluble proteins. Prot Eng Des Sel 22: 67–73

Park Y and Helms V (2007) On the derivation of propensity scales for predicting exposed transmembrane residues of helical membrane proteins. Bioinformatics 23: 701–708

Perozo E, Kloda A, Cortes DM, et al. (2002) Physical principles underlying the transduction of bilayer deformation forces during mechanosensitive channel gating. Nat Struct Biol 9: 696–703

Pitonzo D and Skach WR (2006) Molecular mechanisms of aquaporin biogenesis by the endoplasmic reticulum Sec61 translocon. Biochim Biophys Acta 1758: 976–988

Pitonzo D, Yang Z, Matsumura Y, et al. (2009) Sequence-specific retention and regulated integration of a nascent membrane protein by the endoplasmic reticulum Sec61 translocon. Mol Biol Cell 20: 685–698

Popot JL and Engelman DM (1990) Membrane protein folding and oligomerization: the two-stage model. Biochemistry 29: 4031–4037

Popot JL, Gerchman SE, Engelman DM (1987) Refolding of bacteriorhodopsin in lipid bilayers. A thermodynamically controlled two-stage process. J Mol Biol 198: 655–676

Ren J, Lew S, Wang J, et al. (1999) Control of the transmembrane orientation and interhelical interactions within membranes by hydrophobic helix length. Biochemistry 38: 5905–5912

Ridge KD, Lee SS, Yao LL (1995) In vivo assembly of rhodopsin from expressed polypeptide fragments. Proc Natl Acad Sci USA 92: 3204–3208

Russ WP and Engelman DM (2000) The GxxxG motif: a framework for transmembrane helix–helix association. J Mol Biol 296: 911–919

Sadlish H, Pitonzo D, Johnson AE, et al. (2005) Sequential triage of transmembrane segments by Sec61α during biogenesis of a native multispanning membrane protein. Nat Struct Mol Biol 12: 870–878

Sal-Man N, Gerber D, Bloch I, et al. (2007) Specificity in transmembrane helix–helix interaction mediated by aromatic residues. J Biol Chem 282: 19753–19761

Schneider D and Engelman DM (2004) Motifs of two small residues can assist but are not sufficient to mediate transmembrane helix interactions. J Mol Biol 343: 799–804

Senes A, Engel DE, DeGrado WF (2004) Folding of helical membrane proteins: the role of polar, GxxxG-like and proline motifs. Curr Opin Struct Biol 14: 465–479

Senes A, Ubarretxena-Belandia I, Engelman DM (2001) The Calpha–H ⋯ O hydrogen bond: a determinant of stability and specificity in transmembrane helix interactions. Proc Natl Acad Sci USA 98: 9056–9061

Skach WR (2007) The expanding role of the ER translocon in membrane protein folding. J Cell Biol 179: 1333–1335

Sternberg MJ and Gullick WJ (1990) A sequence motif in the transmembrane region of growth factor receptors with tyrosine kinase activity mediates dimerization. Prot Eng 3: 245–248

Strandberg E, Morein S, Rijkers DTS, et al. (2002) Lipid dependence of membrane anchoring properties and snorkeling behavior of aromatic and charged residues in transmembrane peptides. Biochemistry 41: 7190–7198

Tastan O, Klein-Seetharaman J, Meirovitch H (2009) The effect of loops on the structural organization of a-helical membrane proteins. Biophys J 96: 2299–2312

van Klompenburg W, Nilsson I, von Heijne G, et al. (1997) Anionic phospholipids are determinants of membrane protein topology. EMBO J 16: 4261–4266

von Heijne G (1989) Control of topology and mode of assembly of a polytopic membrane protein by positively charged residues. Nature 341: 456–458

Walters RFS and DeGrado WF (2006) Helix-packing motifs in membrane proteins. Proc Natl Acad Sci USA 103: 13658–13663

White SH and von Heijne G (2008) How translocons select transmembrane helices. Annu Rev Biophys 37: 23–42

Zhang Y, Kulp DW, Lear JD, et al. (2009) Experimental and computational evaluation of forces directing the association of transmembrane helices. J Am Chem Soc 131: 11341–11343

Zhou FX, Cocco MJ, Russ WP, et al. (2000) Interhelical hydrogen bonding drives strong interactions in membrane proteins. Nat Struct Biol 7: 154–160

Prediction of three-dimensional transmembrane helical protein structures

Patrick Barth

Department of Pharmacology, Baylor College of Medicine, One Baylor Plaza, Houston, TX, USA

Abstract

Membrane proteins are critical to living cells and their dysfunction can lead to serious diseases. High-resolution structures of these proteins would provide very valuable information for designing efficient therapies but membrane protein crystallization is a major bottleneck. As an important alternative approach, methods for predicting membrane protein structures have been developed in recent years. This chapter focuses on the problem of modeling the structure of transmembrane helical proteins, and describes recent advancements, current limitations, and future challenges facing de novo modeling, modeling with experimental constraints, and high-resolution comparative modeling of these proteins.

Abbreviations: MP, membrane protein; SP, water-soluble protein; RMSD, root-mean square deviation; $C\alpha$ RMSD, root-mean square deviation over $C\alpha$ atoms; TM, transmembrane; TMH, transmembrane helix; GPCR, G protein-coupled receptor; 3D, three dimensional; NMR, nuclear magnetic resonance spectroscopy; EPR, electron paramagnetic resonance spectroscopy; FTIR, Fourier transform infrared spectroscopy.

1 Introduction

Organization represents one of the key principles of life and is partly achieved, at the cellular level, by compartmentalization where lipid membranes define the boundaries between compartments. Faithful communication across membranes is mostly accomplished by highly specialized membrane proteins (MPs), which transport a wide variety of information from elementary particles to macromolecules or transmit signals through long-range conformational changes. MPs are therefore critical to the regulation of the cell, and the disruption of their functions often has dramatic

Corresponding author: Patrick Barth, Department of Pharmacology, Baylor College of Medicine, One Baylor Plaza, Houston, TX 77030, USA (E-mail: patrickb@bcm.tmc.edu)

effects at cellular and higher levels. Consequently, there is enormous interest in gaining high-resolution structural information on these proteins, which would provide the basis for designing efficient therapies. Despite a few exceptions, X-ray crystallography remains the main method to solve large macromolecular structures at atomic resolution but relies on the ability to organize the molecule regularly in three dimensions. The fluidity of the lipid membrane precludes such organization and led most crystallographers to solubilize MPs in artificial environments, which are often poor mimics of the lipid membranes. Consequently, many MPs resisted crystalization so far leaving us with a database of high-resolution MP structures, which is minimal compared to their soluble counterparts. The three-dimensional (3D) structure prediction of MPs is therefore a very important alternative approach and is the topic of this chapter.

2 Goal of the chapter

MP structures can be classified into two main classes: transmembrane helical MPs and beta-barrel MPs. The problem of predicting the structure of beta-barrel MPs is being discussed in one of the associated chapters. I will exclusively focus on transmembrane helical MP structure prediction. I will preferentially discuss the most recent advancements in the field and try to refer to other reviews for specialized discussions on related topics. Finally, I will discuss limitations of the current methods and future challenges that lie ahead.

3 Methods

To a first approximation, the general principles and strategies underlying MP and water-soluble protein (SP) structure modeling are very similar and are summarized in Fig. 1. All structure modeling tasks typically start with the identification of sequence/structure homologs to the sequence being modeled (i.e., query sequence). The outcome of this search decides whether structure prediction or homology modeling techniques will be used. The thresholds defining structural homologs as well as heuristics and knowledge-based information are different between MPs and SPs and will be discussed in the following sections.

3.1 De novo membrane protein structure prediction

Despite recent progress in the field of MP crystallography (Carpenter et al. 2008), MP structures still represent less than 1% of all protein structures in the Protein Data Bank (PDB; *http://blanco.biomol.uci.edu/Membrane_Proteins_xtal.html*). Consequently, many MP coding sequences are devoid of structural homologs and need to

Query sequence   ALFTLLFSYVVIALTPASVWRALENYDKYLVTALLVIALLSALLTYEANK

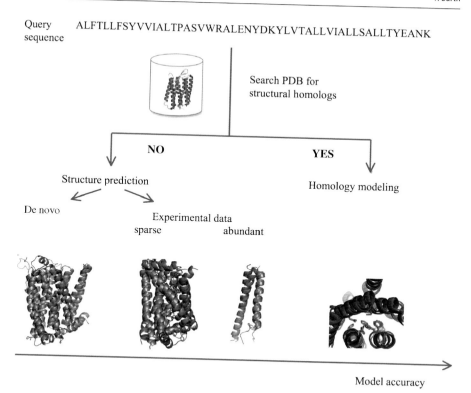

Fig. 1.   General scheme for membrane protein structure modeling. First, bioinformatic techniques are used to search the Protein Data Base (PDB) for structural homologs of the query sequence. If no structural homolog with sequence identity ≥20% is found, the structure is modeled de novo from sequence information alone or by combining structure prediction techniques with diverse experimental data. If structural homologs are found, these are used as templates to build a model for the query sequence by homology modeling techniques. In general, the more experimental information is available for the modeling, the more accurate is the final model.

be modeled from primary sequence information by structure prediction techniques. The MP and SP structure prediction problems differ significantly. On the one hand, MPs are often much larger than the small globular SP domains which are the current target of most protein structure prediction methods (Bradley et al. 2005). Therefore, the conformational space accessible to MP polypeptide chains is extremely large and represents a real search problem for current structure prediction techniques. On the other hand, the lipid membrane environment imposes many constrains on MP structures and, in recent years, strategies have been developed to extract knowledge-based information and define heuristics to restrict the conformational space in folding simulations.

233

Membrane protein folding can be conceptually decomposed into two consecutive steps: folding of the individual hydrophobic segments into helices followed by helix association (Popot and Engelman 1990). Accordingly, the problem of predicting the structure of transmembrane helix (TMH) proteins has been very early on simplified by breaking it down into the following steps: (i) delineating the boundaries of the TMH segments and the topology of the protein (i.e., predicting the membrane embedding of each segment); (ii) predicting the tertiary structure of the protein (i.e., the arrangement and interactions between helices).

3.1.1 MP topology predictions

With the anisotropic constraints imposed by the membrane environment on protein structures and topologies, MP structure prediction becomes a two-body problem where the positions of the protein and the lipid membrane are tightly coupled. The anisotropic physical properties of the environment strongly influence MP sequences and have been used to extract predictive information on the topology and position of MPs in the lipid bilayer. Initially based on simple hydrophobicity scale, topology predictors have recently applied sophisticated machine-learning techniques to extract meaningful amino acid sequence patterns and reliably predict the presence of transmembrane helices (Elofsson and von Heijne 2007). The most accurate methods currently predict the correct topology (i.e., MP orientation in the lipid bilayer and number of transmembrane helices) for up to 89% of all MPs (Tusnády and Simon 2001; Melen et al. 2003; Viklund and Elofsson 2004, 2008; Daley et al. 2005; Kall et al. 2005; Kim et al. 2006; Jones 2007; Bernsel et al. 2008; Viklund et al. 2008; Nugent and Jones 2009). Following their recent work on better understanding peptide recognition by the translocon, Elofsson, Von Heijne, and co-workers developed a topology prediction method based on experimentally determined amino-acid contributions to the free energy of membrane insertion that performs similarly to the best statistically based predictors (Bernsel et al. 2008).

3.1.2 The first MP structure prediction methods developed during the past decade

The first 3D MP structure prediction methods were essentially adaptations of methods initially developed to fold SP 3D structures from primary amino acid sequences (FILM (Pellegrini-Calace et al. 2003) and RosettaMembrane (Yarov-Yarovoy et al. 2006) were adapted from FRAGFOLD (Jones 1997) and Rosetta (Rohl et al. 2004), respectively). The conformational search strategies remained essentially identical and aimed at recapitulating the trade-off between local and non-local interactions during protein folding: short segments of the chain are allowed to alternate between

searching a database of TMH pairs of known structures for local sequence matches with all possible pairs of predicted TMHs in the query sequence (Fig. 3). A library of possible interaction geometries defined by the inter-residue distance and backbone conformations at the interacting positions of the TMH pair from the database is generated for each pair of predicted helices in the query. During folding simulations, predicted helix pairs are constrained with a single randomly selected predicted interaction in the library. To compensate for the low accuracy of the predicted interactions (around 30% were native-like), a minimum of ten predicted interactions are tested for each helix pair, which allowed correct models to be generated for several MPs with complex topologies (Barth et al. 2009).

3.1.3.2 Contact predictors

In addition to the above-mentioned sequence/structure correlations, other studies identified specific residue motifs stabilizing helix–helix interactions (i.e., polar residues through formation of hydrogen bonds (Zhou et al. 2001), the aromatic–XXaromatic motif (Sal-Man et al. 2007), the GXXXG motif found in glycophorin A (Lemmon et al. 1992) and heptad motifs of leucine residues (Gurezka et al. 1999). The recurrent observation of such generic helix–helix stabilizing motifs strongly suggests their predictability from sequence by generalized pattern search strategy.

Several groups have used machine-learning techniques to develop residue contact predictors specifically trained on MP structures. Among the two best performing predictors, TMhit applies a support vector machine (SVM) classifier within a hierarchical framework, which first identifies potential contacting residues on a per residue basis and then predicts the contact patterns from all possible pairs of contacting residues (Lo et al. 2009). Another method, TMHcon (Fuchs et al. 2009) combines residue co-evolution information, residue position within the TM helix, predicted lipid exposure using the LIPS method (Adamian and Liang 2006), with a neural network and profile data, in order to predict helix–helix interaction. TMHcon and TMhit were shown to perform close or better than the best contact predictors developed for SPs with contact prediction accuracies of nearly 26% and 31%, respectively. When three unique contact pairs define interacting helical pairs, TMhit predicted such pairs with accuracy, sensitivity, and specificity of 56%, 40%, and 89%, respectively. As the number of contact defining an interacting helical pair increases, accuracy and specificity increase at the expense of sensitivity. It has been reported that near-native structures of SP can be recapitulated with an average of one contact for every eight residues (Li et al. 2004). The recent results obtained by RosettaMembrane suggest that this number might be lower for MPs and that libraries of predicted contacts could be directly used as constraints to fold large MP structures.

Although the results obtained with TMHhit and TMHcon are promising, the dataset of MP structures is still relatively small and loose homology cutoff has sometimes been adopted to generate the dataset for training the machine-learning approaches. As structural homologs can be found that have lower than 20% sequence identity, it will be important in future studies to train such predictors with stringent sequence identity thresholds.

3.1.4 MP-specific energy functions for decoy discrimination

In blind protein structure prediction applications, the ability to discriminate by energy near-native from non-native models is critical and has led to several developments in the field of energy functions for MP structures. Because modeling explicitly all the components of lipid membrane bilayers is computationally very expensive, most energy functions treat the lipid bilayer implicitly as an anisotropic solvent. Pure knowledge-based potentials, such as Ez (Senes et al. 2007), were developed to recapitulate the energies of amino acid insertions in the membrane. These potentials are derived from the distribution of amino acids at particular positions in MP structures, i.e., at particular depth in the membrane. Remarkably, the Ez potential demonstrates good correlations with the in vivo energy scale of membrane peptide insertions mediated by the translocon (Hessa et al. 2005). However, although an atom-based version has been discussed for Ez, the knowledge-based potentials are essentially residue-based and do not treat specific solvation effects associated with different side-chain conformations.

Implicit atomic solvation potentials have been developed based on experimental free energy of transfer of amino acid analogs from vacuum to organic solvents (Lazaridis and Karplus 2003; Barth et al. 2007). These potentials model the membrane with a three-phase system: two isotropic phases for water and the hydrophobic core of the membrane and one anisotropic phase in between which interpolates the properties of the adjacent phases. More rigorous physically based descriptions of the membrane based on Poisson–Boltzmann or Generalized-Born formalisms have also been developed and treat the anisotropy of the membrane with multiple layers of different dielectric properties (Roux 2002; Im et al. 2005; Tanizaki and Feig 2005; Feig 2008).

In addition to the membrane-specific solvation energies, most above-mentioned atom-based potentials model inter-atomic interactions (e.g., Van der Waals and hydrogen-bonding) explicitly following the formalisms developed for Molecular Mechanics force fields. Although IMM1 treats weak CαH–O hydrogen bonds as pure electrostatic interactions, this level of description was sufficient to reconcile apparently contradicting experimental results on the role of these bonds in MP stability (Mottamal and Lazaridis 2005). Unlike Molecular Mechanics force field, Rosetta models the orientation dependencies of hydrogen-bond energies resulting from their partial covalent character. The application of such potential in docking and design calculations demonstrated that weak $C\alpha H$–O hydrogen bonds can contribute significantly to the stability and specificity of TMH–TMH interactions (Barth et al. 2007).

Few of these potentials have been used in structure prediction calculations. IMM1 was shown to recapitulate reasonably well the interactions of peptides with lipid charged headgroup regions and mutational effects on MP stability (Lazaridis 2005; Mottamal and Lazaridis 2005). However, when applied to the prediction of the dimeric structure of glycophorin A, IMM1 was not able to select by energy the native from non-native topologies (Mottamal et al. 2006). By combining replica exchange MD simulations with a Generalized-Born implicit membrane model, Brooks et al. attempted to predict the de novo structures of several simple helix homo-oligomers. When the native oligomerization state was enforced, representative models from the largest families were identified that closely matched the experimentally determined structures for the dimeric glycophorin A, the M2 proton channel, and phospholamban (Bu et al. 2007). However, the native oligomeric state could not always be selected by energy from alternative topologies. RosettaMembrane was recently applied to the structure prediction of several integral MP ranging from domains of 4 TMHs to full-length MPs of 7 TMHs. The atomic potential was used to refine coarse-grained models and when such models were within 4 Å of the X-ray structures, they were often within the top 5 lowest energy structures after all-atom refinement (Barth et al. 2007, 2009).

Further validation of these potentials in blind prediction tests will be necessary but one can already point at one main limitation in the implicit description of the membrane. All these models assume a planar rigid and symmetric lipid bilayer. However, many evidences suggest that the membrane itself can adapt to protein conformations and achieve lower energy protein/lipid configurations (Bowie 2005; Hessa et al. 2005). Future research in this direction will likely involve the implementation of membrane deformation properties observed in simulations with explicit lipids within implicit membrane models (Dorairaj and Allen 2007; Choe et al. 2008).

3.2 Sequence-based modeling with experimental constraints

Although high-resolution atomic structures are not yet available for many MPs, experimental studies have generated a large diversity of data inferring residue contacts or proximity in MP structures. Such experimental information can often be used to restrict the conformational space of the polypeptide during folding simulations or to filter the models afterwards, therefore improving the quality of the models. These data involve experimental techniques such as site mutagenesis, chemical cross-linking, NMR, EPR, FTIR as well as low- to medium-resolution structures. The studies described below are not meant to be exhaustive but representative examples of different combinations of these techniques with structure prediction methods.

Compensatory mutations, i.e., when the effect of a point mutation at one site can be compensated by another mutation at another site, can infer a physical proximity between the two sites. For example, compensatory mutations suggested that Asp 237 and Lys 358 interact directly via a salt bridge in the core of the C-terminal domain of lactose permease (Zhao et al. 1999), a conclusion that was later confirmed by the X-ray structure (Abramson et al. 2003). Although charged residues can adopt many conformations to form a salt bridge, this low-resolution information was sufficient to constrain the distance and position of the two residues and generate near-native models of the domain (Barth et al. 2009).

Disulfide cross-linking experiments have been very popular over the last two decades and can probe residue proximity and direct contacts in protein structures through the formation of covalent disulfide bonds. This technique has been used extensively to probe the topology and domain organization of the lactose permease and bacterial chemoreceptors for example (Wu and Kaback 1996; Wu et al. 1996; Bass et al. 2007). It was recently combined with Rosetta to describe the 3D structure at atomic level of the intact integrin receptor on the cell surface (Zhu et al. 2009). Individual positions in the transmembrane and juxtamembrane of each monomer of the αIIbβ3 receptors were mutated to cysteines and the ability of each possible cysteine pairs to cross-link was monitored in native membrane at the cell surface. Cross-linking patterns were translated into 48 distance constraints and combined with RosettaMembrane to model the domains. The final structure of the TM and juxtamembrane showed remarkable structural similarity (RMSD: 2.1 Å) with a structure concurrently solved by high-resolution solid-state NMR techniques (Lau et al. 2009). Because the formation of covalent disulfide bonds can disturb the structures of flexible regions and/or stabilize protein conformations that are rarely occupied at equilibrium, these data have to be carefully interpreted in structure modeling applications.

The presence of cofactors imposes stringent constraints on protein structures and can be used as another potential source of restraints in structure predictions. Cofactors are ubiquitously involved in the catalytic activities of enzymes and have structurally well-defined chemical structures. Protein residues chelating these cofactors have well-defined chemical properties, are evolutionarily conserved, and can often be identified by sequence information alone. The heme-binding subunits of fumarate reductase and cytochrome bc1 were predicted by modeling the orientations between the two heme-chelating histidines with a library of helical pairs binding hemes extracted from the PDB. These sparse constraints were sufficient to predict the native topology of these two protein domains (Barth et al. 2009).

When experimental data are available from different sources, they can be combined to generate a reasonably sized library of distance constraints to model the structure of large MP structures. An early demonstration of such approach involved the modeling of bovine rhodopsin with 27 distance restraints from EPR, FTIR, and chemical cross-linking experiments (Sale et al. 2004). The final model was near-native with a root-mean square derivation over Cα atom (Cα RMSD) of only 3.2 Å to the X-ray structure.

Solid-state NMR techniques have gained in resolution in recent years and observables such as ^{15}N chemical shift and $^{15}N-^{1}H$ dipolar coupling can now be used in conjunction with modeling techniques to determine the structure of oligomers and their orientation in the lipid bilayer (Lee et al. 2008).

Finally, low- to medium-resolution structural information from cryo-electron microscopy (cryo-EM) can provide starting information to restrict the number of possible TMH arrangements and TMH orientations in the membrane prior to structure modeling efforts. The resolution of the most accurate cryo-EM structures in the plane of the membrane typically ranges from 5 to 10 Å, allowing helical axis to be defined but precluding the observation of amino acids and therefore the assignment of TMHs. The cryo-EM data can be supplemented with sequence conservation, hydrophobicity patterns, mutagenesis data, and length of connecting loops between TMHs. Then, the modeling consists in assigning the helices, predicting the orientation of the helices around their axis, and building connecting loops (see Fleishman and Ben-Tal 2006; Topf et al. 2008; for a detailed description of the specific methods developed for the modeling).

In recent years, main efforts combining de novo structure modeling with cryo-EM data have targeted for example the small-drug transporter EmrE (Fleishman et al. 2006), the gap-junction intercellular channel (Fleishman et al. 2004), and the oxalate transporter (Beuming and Weinstein 2005). The models generated for EmrE proved to be close to the native topology after a long controversy led by wrong X-ray structures was solved with the recent release of the correct X-ray structure (Korkhov and Tate 2009). The model based on cryo-EM data was found to be very similar to the revised X-ray structure with a Cα RMSD of only 1.4 Å (Korkhov and Tate 2009). In the case of the Gap-junction, however, the coarse-grained model based on cryo-EM and interpretation of experimental data differs significantly from the recently solved X-ray structure. These discrepancies point toward the difficulty of combining and interpreting unambiguously experimental data from different sources when assigning TMH (Maeda et al. 2009).

As several techniques are currently developed to characterize the structure at low-resolution of MPs in physiologically-relevant environments (Bartesaghi and Subramaniam 2009), the combination of modeling and intermediate-resolution structure determination techniques offers great promises for the future as powerful complementary approaches to high-resolution X-ray crystallography.

3.3 Comparative modeling of MP structures

Proteins with similar sequences often adopt similar structures. This simple observation led to the development of comparative modeling techniques, which have been widely applied to SP structure modeling (Baker and Sali 2001). The recent progress in MP structure determination and the expected restricted structural space accessible to TMH proteins hold promises that such technique may become soon an important approach for high-resolution MP structure modeling.

A study analyzed carefully the applicability of the methods originally developed for SP to MP structure modeling (Forrest et al. 2006). Secondary structure prediction methods like PSIPRED were found to perform similarly for both SP and TMH proteins. Accurate sequence alignments could be obtained with the best profile–profile alignment techniques developed for SPs. Finally, a sequence identity threshold of 30% combined with an accurate sequence alignment was found to provide reasonably accurate models in the TMH regions (Cα RMSD ≤2 Å to the native structure).

Over the past few years, many modeling studies have taken advantage of MP families sharing similar folds (e.g., GPCR or ion channels) to generate structural models of important targets and interpret functional and mutational data. However, the first stringent test of existing methods only occurred during fall 2008 with the critical assessment of GPCR structure modeling, organized for the blind prediction of the adenosine A_{2A} receptor structure (Michino et al. 2009). The closest structural homologs, the beta adrenergic receptors, share ~30% sequence identity with the adenosine receptor. In the aligned helical regions, the beta2 adrenergic receptor structure has a Cα RMSD of 2.8 Å to the adenosine receptor structure. A total of 206 models were submitted for which an average RMSD of 2.8 Å to the adenosine receptor in the helical regions was reported. Very few models showed significant improvements over the template in the helical regions (Fig. 4) and were generated with modeling techniques combining multiple templates (Michino et al. 2009) or involving all-atom refinement specifically designed for TMH proteins such as RosettaMembrane (Barth et al. 2007). As for SPs, these results clearly demonstrate the difficulty of refining structural models below 2 Å with current modeling techniques.

Most contacts between the GPCR and the ligands structurally characterized so far involve amino acids in the TMH region of the receptor and were relatively well predicted in the best models of the adenosine receptor. However, two residues of the long disordered extracellular loop (ECL2), which shares very low sequence identity with the adrenergic receptors, make also crucial contacts with the ligand. The ECL2

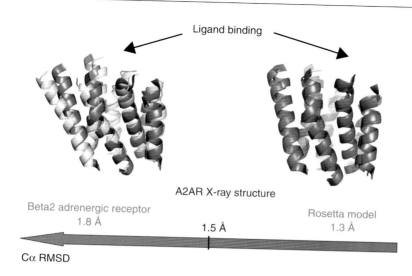

Fig. 4.   **Accurate modeling of the TM helical shifts shaping the ligand-binding site in the blind prediction of the adenosine receptor structure (Michino et al. 2009).** The starting template, beta2 adrenergic receptor structure, has a 1.8 Å Cα RMSD to the native in the TMH region. The comparative modeling mode of RosettaMembrane was able to improve the starting template in this region by 0.5 Å.

loop was often rebuilt de novo and poorly modeled when compared to the X-ray structure. Consequently, the ligand position was not predicted accurately and exhibited RMSD over the ligand atoms and a percent of native contacts to the receptor ranging from 2.6 to 6.7 Å and 53 to 4% in the top 10 predictions, respectively.

Modeling techniques currently emerge that can significantly and consistently improve starting templates in the TMH regions (Barth P, personal communication). However, the main future challenge lies in the modeling of long partially disordered loops, which can play critical role in ligand binding.

4 Conclusions and future directions

Over the past decade, the number of MP structures solved by X-ray crystallography has increased significantly to a point where knowledge-based structure prediction techniques can be developed and tested. The SP structure prediction field has greatly benefited from the organization of blind predictions with critical assessment of protein structure prediction (CASP). Similarly, it is expected that such stringent tests of existing structure prediction methods for MPs will help better define directions for improvements. In that respect, the first blind prediction of a GPCR structure was very informative for the comparative modeling field and we look forward to similar

CASPs for other MPs in the future or to the publication of blind predicted models (Zhang et al. 2006; Bu and Brooks 2008).

How MP structures help us understand their function in native membranes is a fundamental question, which I have voluntarily overlooked and would probably necessitate a chapter on its own. Much experimental evidence suggests that conformational change is one hallmark of MP function and regulation. X-ray structures may only represent a snapshot in the conformational space accessible to MPs in their native environment. One of the main future challenges will be to reliably model alternative states of MPs and stringently validate these predictions, a problem where the quality of both computational techniques and experimental structural data will matter.

Note added in proof: A new method for predicting helix packing arrangement using a contact predictor has been published recently: Nugent and Jones (2010) Plos Comp Biol Mar 19: 6(3) e1000714.

References

Abramson J, Smirnova I, Kasho V, Verner G, Kaback HR, Iwata S (2003) Structure and mechanism of the lactose permease of *Escherichia coli*. Science 301(5633): 610–615

Adamian L and Liang J (2006) Prediction of transmembrane helix orientation in polytopic membrane proteins. BMC Struct Biol 6: 13

Baker D and Sali A (2001) Protein structure prediction and structural genomics. Science 294(5540): 93–96

Bartesaghi A and Subramaniam S (2009) Membrane protein structure determination using cryo-electron tomography and 3D image averaging. Curr Opin Struct Biol 19(4): 402–407

Barth P, Schonbrun J, Baker D (2007) Toward high-resolution prediction and design of transmembrane helical protein structures. Proc Natl Acad Sci USA 104(40): 15682–15687

Barth P, Wallner B, Baker D (2009) Prediction of membrane protein structures with complex topologies using limited constraints. Proc Natl Acad Sci USA 106(5): 1409–1414

Bass RB, Butler SL, Chervitz SA, Gloor SL, Falke JJ (2007) Use of site-directed cysteine and disulfide chemistry to probe protein structure and dynamics: applications to soluble and transmembrane receptors of bacterial chemotaxis. Meth Enzymol 423: 25–51

Becker OM, Marantz Y, Shacham S, Inbal B, Heifetz A, Kalid O, et al. (2004) G protein-coupled receptors: in silico drug discovery in 3D. Proc Natl Acad Sci USA 101(31): 11304–11309

Bernsel A, Viklund H, Falk J, Lindahl E, von Heijne G, Elofsson A (2008) Prediction of membrane-protein topology from first principles. Proc Natl Acad Sci USA 105(20): 7177–7181

Beuming T and Weinstein H (2005) Modeling membrane proteins based on low-resolution electron microscopy maps: a template for the TM domains of the oxalate transporter OxlT. Protein Eng Des Sel 18(3): 119–125

Bowie JU (2005) Solving the membrane protein folding problem. Nature 438(7068): 581–589

Bradley P, Misura KM, Baker D (2005) Toward high-resolution de novo structure prediction for small proteins. Science 309(5742): 1868–1871

Bu L and Brooks CL III (2008) De novo prediction of the structures of *M. tuberculosis* membrane proteins. J Am Chem Soc 130(16): 5384–5385

Bu L, Im W, Brooks CL III (2007) Membrane assembly of simple helix homo-oligomers studied via molecular dynamics simulations. Biophys J 92(3): 854–863

Carpenter EP, Beis K, Cameron AD, Iwata S (2008) Overcoming the challenges of membrane protein crystallography. Curr Opin Struct Biol 18(5): 581–586

Cherezov V, Rosenbaum DM, Hanson MA, Rasmussen SG, Thian FS, Kobilka TS, et al. (2007) High-resolution crystal structure of an engineered human beta2-adrenergic G protein-coupled receptor. Science 318(5854): 1258–1265

Choe S, Hecht KA, Grabe M (2008) A continuum method for determining membrane protein insertion energies and the problem of charged residues. J Gen Physiol 131(6): 563–573

Daley DO, Rapp M, Granseth E, Melen K, Drew D, von Heijne G (2005) Global topology analysis of the *Escherichia coli* inner membrane proteome. Science 308(5726): 1321–1323

Dorairaj S and Allen TW (2007) On the thermodynamic stability of a charged arginine side chain in a transmembrane helix. Proc Natl Acad Sci USA 104(12): 4943–4948

Elofsson A and von Heijne G (2007) Membrane protein structure: prediction versus reality. Annu Rev Biochem 76: 125–140

Feig M (2008) Implicit membrane models for membrane protein simulation. Method Mol Biol 443: 181–196

Fleishman SJ and Ben-Tal N (2006) Progress in structure prediction of alpha-helical membrane proteins. Curr Opin Struct Biol 16(4): 496–504

Fleishman SJ, Unger VM, Yeager M, Ben-Tal N (2004) A Calpha model for the transmembrane alpha helices of gap junction intercellular channels. Mol Cell 15(6): 879–888

Fleishman SJ, Harrington SE, Enosh A, Halperin D, Tate CG, Ben-Tal N (2006) Quasi-symmetry in the cryo-EM structure of EmrE provides the key to modeling its transmembrane domain. J Mol Biol 364(1): 54–67

Forrest LR, Tang CL, Honig B (2006) On the accuracy of homology modeling and sequence alignment methods applied to membrane proteins. Biophys J 91(2): 508–517

Fuchs A, Kirschner A, Frishman D (2009) Prediction of helix–helix contacts and interacting helices in polytopic membrane proteins using neural networks. Proteins 74(4): 857–871

Gurezka R, Laage R, Brosig B, Langosch D (1999) A heptad motif of leucine residues found in membrane proteins can drive self-assembly of artificial transmembrane segments. J Biol Chem 274(14): 9265–9270

Hessa T, Kim H, Bihlmaier K, Lundin C, Boekel J, Andersson H, et al. (2005) Recognition of transmembrane helices by the endoplasmic reticulum translocon. Nature 433(7024): 377–381

Im W and Brooks CL III (2005) Interfacial folding and membrane insertion of designed peptides studied by molecular dynamics simulations. Proc Natl Acad Sci USA 102(19): 6771–6776

Jaakola VP, Griffith MT, Hanson MA, Cherezov V, Chien EY, Lane JR, et al. (2008) The 2.6 Angstrom crystal structure of a human A2A adenosine receptor bound to an antagonist. Science 322(5905): 1211–1217

Jones DT (1997) Successful ab initio prediction of the tertiary structure of NK-lysin using multiple sequences and recognized supersecondary structural motifs. Proteins (Suppl 1): 185–191

Jones DT (2007) Improving the accuracy of transmembrane protein topology prediction using evolutionary information. Bioinformatics 23(5): 538–544

Kall L, Krogh A, Sonnhammer EL (2005) An HMM posterior decoder for sequence feature prediction that includes homology information. Bioinformatics 21 (Suppl 1): i251–i257

Kauko A, Illergard K, Elofsson A (2008) Coils in the membrane core are conserved and functionally important. J Mol Biol 380(1): 170–180

Kim H, Melen K, Osterberg M, von Heijne G (2006) A global topology map of the *Saccharomyces cerevisiae* membrane proteome. Proc Natl Acad Sci USA 103(30): 11142–11147

Korkhov VM and Tate CG. An emerging consensus for the structure of EmrE. Acta Crystallogr D Biol Crystallogr 65(Pt 2): 186–192

Lau TL, Kim C, Ginsberg MH, Ulmer TS (2009) The structure of the integrin alphaIIbbeta3 transmembrane complex explains integrin transmembrane signalling. EMBO J 28(9): 1351–1361

Lazaridis T (2005) Implicit solvent simulations of peptide interactions with anionic lipid membranes. Proteins 58(3): 518–527

Lazaridis T and Karplus M (2003) Thermodynamics of protein folding: a microscopic view. Biophys Chem 100(1–3): 367–395

Lee J, Chen J, Brooks CL III, Im W (2008) Application of solid-state NMR restraint potentials in membrane protein modeling. J Magn Reson 193(1): 68–76

Lemmon MA, Flanagan JM, Hunt JF, Adair BD, Bormann BJ, Dempsey CE, et al. (1992) Glycophorin A dimerization is driven by specific interactions between transmembrane alpha-helices. J Biol Chem 267(11): 7683–7689

Li W, Zhang Y, Skolnick J (2004) Application of sparse NMR restraints to large-scale protein structure prediction. Biophys J 87(2): 1241–1248

Li Y and Goddard WA III (2008) Prediction of structure of G-protein coupled receptors and of bound ligands, with applications for drug design. Pac Symp Biocomput: 344–353

Lo A, Chiu YY, Rodland EA, Lyu PC, Sung TY, Hsu WL (2009) Predicting helix–helix interactions from residue contacts in membrane proteins. Bioinformatics 25(8): 996–1003

Maeda S, Nakagawa S, Suga M, Yamashita E, Oshima A, Fujiyoshi Y, et al. (2009) Structure of the connexin 26 gap junction channel at 3.5 A resolution. Nature 458(7238): 597–602

Melen K, Krogh A, von Heijne G (2003) Reliability measures for membrane protein topology prediction algorithms. J Mol Biol 327(3): 735–744

Michino M, Abola E, Brooks CL III, Dixon JS, Moult J, Stevens RC (2009) Community-wide assessment of GPCR structure modelling and ligand docking: GPCR Dock 2008. Nat Rev Drug Discov 8(6): 455–463

Mottamal M and Lazaridis T (2005) The contribution of C alpha–H⋯O hydrogen bonds to membrane protein stability depends on the position of the amide. Biochemistry 44(5): 1607–1613

Mottamal M, Zhang J, Lazaridis T (2006) Energetics of the native and non-native states of the glycophorin transmembrane helix dimer. Proteins 62(4): 996–1009

Nugent T and Jones DT (2009) Transmembrane protein topology prediction using support vector machines. BMC Bioinform 10: 159

Pellegrini-Calace M, Carotti A, Jones DT (2003) Folding in lipid membranes (FILM): a novel method for the prediction of small membrane protein 3D structures. Proteins 50(4): 537–545

Popot JL and Engelman DM (1990) Membrane protein folding and oligomerization: the two-stage model. Biochemistry 29(17): 4031–4037

Rohl CA, Strauss CE, Misura KM, Baker D (2004) Protein structure prediction using Rosetta. Method Enzymol 383: 66–93

Roux B (2002) Theoretical and computational models of ion channels. Curr Opin Struct Biol 12(2): 182–189

Sale K, Faulon JL, Gray GA, Schoeniger JS, Young MM (2004) Optimal bundling of transmembrane helices using sparse distance constraints. Protein Sci 13(10): 2613–2627

Sal-Man N, Gerber D, Bloch I, Shai Y (2007) Specificity in transmembrane helix–helix interactions mediated by aromatic residues. J Biol Chem 282(27): 19753–19761

Senes A, Chadi DC, Law PB, Walters RF, Nanda V, Degrado WF (2007) E(z), a depth-dependent potential for assessing the energies of insertion of amino acid side-chains into membranes: derivation and applications to determining the orientation of transmembrane and interfacial helices. J Mol Biol 366(2): 436–448

Shacham S, Marantz Y, Bar-Haim S, Kalid O, Warshaviak D, Avisar N, et al. (2004) PREDICT modeling and in-silico screening for G-protein coupled receptors. Proteins 57(1): 51–86

Simons KT, Kooperberg C, Huang E, Baker D (1997) Assembly of protein tertiary structures from fragments with similar local sequences using simulated annealing and Bayesian scoring functions. J Mol Biol 268(1): 209–225

Tanizaki S and Feig M (2005) A generalized Born formalism for heterogeneous dielectric environments: application to the implicit modeling of biological membranes. J Chem Phys 122(12): 124706

Topf M, Lasker K, Webb B, Wolfson H, Chiu W, Sali A (2008) Protein structure fitting and refinement guided by cryo-EM density. Structure 16(2): 295–307

Tusnády GE and Simon I (2001) The HMMTOP transmembrane topology prediction server. Bioinformatics 17(9): 849–850

Vaidehi N, Floriano WB, Trabanino R, Hall SE, Freddolino P, Choi EJ, et al. (2002) Prediction of structure and function of G protein-coupled receptors. Proc Natl Acad Sci USA 99(20): 12622–12627

Viklund H and Elofsson A (2004) Best alpha-helical transmembrane protein topology predictions are achieved using hidden Markov models and evolutionary information. Protein Sci 13(7): 1908–1917

Viklund H and Elofsson A (2008) OCTOPUS: improving topology prediction by two-track ANN-based preference scores and an extended topological grammar. Bioinformatics 24(15): 1662–1668

Viklund H, Bernsel A, Skwark M, Elofsson A (2008) SPOCTOPUS: a combined predictor of signal peptides and membrane protein topology. Bioinformatics 24(24): 2928–2929

Walters RF and DeGrado WF (2006) Helix-packing motifs in membrane proteins. Proc Natl Acad Sci USA 103(37): 13658–13663

Warne T, Serrano-Vega MJ, Baker JG, Moukhametzianov R, Edwards PC, Henderson R, et al. (2008) Structure of a beta1-adrenergic G-protein-coupled receptor. Nature 454(7203): 486–491

Wu J and Kaback HR (1996) A general method for determining helix packing in membrane proteins in situ: helices I and II are close to helix VII in the lactose permease of *Escherichia coli*. Proc Natl Acad Sci USA 93(25): 14498–14502

Wu J, Voss J, Hubbell WL, Kaback HR (1996) Site-directed spin labeling and chemical crosslinking demonstrate that helix V is close to helices VII and VIII in the lactose permease of *Escherichia coli*. Proc Natl Acad Sci USA 93(19): 10123–10127

Yarov-Yarovoy V, Schonbrun J, Baker D (2006) Multipass membrane protein structure prediction using Rosetta. Proteins 62(4): 1010–1025

Zhang Y, Devries ME, Skolnick J (2006) Structure modeling of all identified G protein-coupled receptors in the human genome. PLoS Comput Biol 2(2): e13

Zhao M, Zen KC, Hubbell WL, Kaback HR (1999) Proximity between Glu126 and Arg144 in the lactose permease of *Escherichia coli*. Biochemistry 38(23): 7407–7412

Zhou FX, Merianos HJ, Brunger AT, Engelman DM (2001) Polar residues drive association of polyleucine transmembrane helices. Proc Natl Acad Sci USA 98(5): 2250–2255

Zhu J, Luo BH, Barth P, Schonbrun J, Baker D, Springer TA (2009) The structure of a receptor with two associating transmembrane domains on the cell surface: integrin alphaIIbbeta3. Mol Cell 34(2): 234–249

GPCRs: past, present, and future

Bas Vroling, Robert P. Bywater, Laerte Oliveira and Gert Vriend

Medical Centre, CMBI, NCMLS, Radboud University Nijmegen, Nijmegen, The Netherlands

Abstract

The family of G-protein-coupled receptors (GPCRs) is by far the best-studied family among the integral membrane proteins, because it represents the largest and most important group for therapeutics. In this chapter we provide an overview of the major developments in the GPCR field since the 19th century, and we shed some light on some of the questions that are relevant now and those that need to be answered in the future regarding GPCR structure and function.

1 Introduction

Over 1000 human genes encode G-protein-coupled receptors (GPCRs). The ligands that bind or otherwise activate these receptors are heterogeneous and include photons, odors, pheromones, hormones, ions, neurotransmitters, and proteases. GPCRs transmit signals from outside the cell to amplification cascades controlling sight, taste, smell, slow neurotransmission, cell division, etc.

GPCRs were long thought to perform a relatively straightforward role; coupling the binding of agonists to the activation of G-proteins, which in turn leads modulation of other downstream effector proteins. However, in recent years it has become clear that many GPCRs have much more complex signaling characteristics. Many GPCRs are constitutively active, and this allows for a fine-grained control of the amount of G-protein activation, being subject to regulation by agonist as well as inverse agonists (Bond and IJzerman 2006). Signaling can occur by using multiple subtypes of G-proteins, or without using G-proteins altogether (Ritter and Hall 2009). Desensitization processes can involve multiple pathways, including phosphorylation events, arrestin-mediated receptor internalization, receptor recycling, and lysosomal

Corresponding author: Bas Vroling, Medical Centre, NCMLS, Radboud University Nijmegen, 6525 GA Nijmegen, The Netherlands (E-mail: bvroling@cmbi.ru.nl)

degradation (Hanyaloglu and von Zastrow 2008; Tobin 2008; Kovacs et al. 2009). The reason for this multifaceted behavior may be the fact that, while there are only about 1000 GPCRs that can be activated by an even smaller number of endogenous agonists, these receptors need to cater for many 1000s of different messages that the whole organism needs to be able to transmit internally. The exact mechanism of this switching between different functions is unknown.

All GPCRs form a bundle of seven transmembrane (TM) helices, connected by three intracellular and three extracellular loops. Although sequence similarities within a single family can be lower even than 25%, there are a number of conserved sequence motifs that imply shared structural features and activation mechanisms.

GPCRs are a major target for the pharmaceutical industry as is reflected by the fact that more than a quarter of all FDA approved drugs act on a GPCR (Overington et al. 2006). Despite intensive academic and industrial research efforts over the past three decades, little is known about the structural basis of GPCR function, in particular, the switches between different functions, referred to above, and between the active and inactive states of these receptors (Rognan 2006).

Some of the major questions relevant to fundamental research into GPCR pharmacology include the following: What residues are critical for ligand binding and for the activation of G-proteins or other proteins? What do different receptor families have in common with regard to their activation mechanism? And which residues are responsible for the differences and should thus especially, or especially not be influenced by potential drug molecules?

The GPCR field was without new structural information for almost a decade, but a number of high-resolution crystal structures have become available recently, giving the GPCR field a big stimulus. As is often the case in such situations, researchers mainly focused on the unique and exciting aspects of these structures, tending to overstate the relevance and importance of the differences. The new structures provide us with new insights, but they are not the holy grail of structural biology of the GPCR field, and many questions, limitations, and challenges remain.

2 A short history

Since the beginning of the 19th century, pharmacologists have studied the dose-dependent effects of neurotransmitters, peptides, and other chemicals on tissues, organs, and animal models. Langley and Dale were the first who explicitly stated the idea of a "receptive substance" on reactive cells (Langley 1905). They performed their experiments on muscle preparations and salivary glands. The targets that they were investigating later turned out to be GPCRs and ion channels. During the next 50 years, the elementary concepts of Langley and Dale were developed into the receptor theory by using physiologic techniques to study receptors (Maehle 2004).

The field of receptor studies was mainly a pharmacological field until – in the 1960s and early 1970s – biochemists became involved in the search for the molecular basis of hormone and drug action. This led to the rapid discovery of the elements that make up the signaling cascades that couple hormones to the intracellular effector proteins. Sutherland discovered the enzyme adenylyl cyclase (Rall and Sutherland 1958), which is responsible for the synthesis of cAMP, and cAMP itself (Sutherland and Rall 1958) that mediates the actions of many receptors. Around the same time, Krebs discovered the cAMP-dependent protein kinase (Walsh et al. 1968) and Gilman had demonstrated the existence of a protein, Gs, that functioned as a transducer between hormone receptors and adenylyl cyclase (Ross and Gilman 1977; Gilman 1987).

The impact of these discoveries was enormous; for the first time scientists could study receptors at a level much closer to the actual signaling than was possible ever before. No longer was it necessary to measure complex physiologic responses such as muscle contraction or gland secretion; now it was possible to measure direct downstream effects such as secondary messenger generation.

After the discovery of downstream effectors, researchers were searching for an even more direct measure to study receptors. There was a need for a means of identifying and studying receptors directly, so that their properties no longer needed to be inferred from downstream effects. Radioligand-binding methods provided these means, and the development of radioligand-binding methods during the 1970s transformed the field of receptor research (Rodbell et al. 1971; Pert and Snyder 1973; Yamamura and Snyder 1974; Mukherjee et al. 1975). By using these techniques, it was now possible to develop approaches to analyze receptor interactions with G-proteins (De Lean et al. 1980). This ultimately led to the development of the "ternary complex model", which provided a way of quantifying coupling efficiency of the receptors to the G-proteins (Mickey et al. 1975; Kent et al. 1980).

Radioligand-binding methods also allowed for new types of ligand-binding studies. New chemical compounds could now be tested systematically, and binding profiles were generated on membrane preparations from different tissues. This led to the first evidence for receptor subtypes expressed in different tissues.

The development of radioligand-binding methods was not only of great importance for the study of receptor properties in native environments. New technologies that built on the principles of molecular recognition were introduced, such as radioligand techniques for determining ligand binding and activity, and affinity chromatography, which allowed for the creation of enriched and purified sources of receptors. In 1979, the Lefkowitz group was able to purify the β2-adrenoceptor by using a broad range of already available β2-adrenergic ligands and a combination of already existing and new affinity chromatography procedures (Caron et al. 1979). Receptors were reconstituted in phospholipid vesicles with purified G-protein and the catalytic moiety of adenylate cyclase, thereby proving that the purified β2-adrenoceptor was indeed the functional receptor (Cerione et al. 1983, 1984; May et al. 1985).

In 1982, Ovchinnikov determined the amino acid sequence of rhodopsin using classical protein sequencing techniques. The amount of protein needed for these techniques was relatively large, but the easy access to retinal rod preparations allowed Ovchinnikov to obtain the necessary amounts of receptor needed for the sequencing. An extensive study that included a combination of bioinformatics and biochemistry showed that rhodopsin had a 7TM architecture and the fact that it had this in common with bacteriorhodopsin was quickly noted (Ovchinnikov YuA 1982; Hargrave et al. 1983). Bacteriorhodopsin is a photon-driven retinal-binding proton pump for which, Henderson and Unwin (1975) had determined a 7TM topology using electron microscopy techniques. After the discovery that bacteriorhodopsin as well as rhodopsin had a 7TM architecture, it was concluded that this 7TM topology was a common feature of light-sensitive proteins (Ovchinnikov YuA 1982).

In 1986 came an important breakthrough: the cloning of the hamster β2-adreno-ceptor (Dixon et al. 1986). The gene for the β2-adrenoceptor was intronless, a feature that is also seen in many other GPCR family members (Kobilka et al. 1987b,c; Sunahara et al. 1990; Pepitoni et al. 1997). The sequence of the β2-adrenoceptor revealed that this receptor shared sequence similarity and a predicted 7TM topology with rhodopsin (Dixon et al. 1986). At the time it was well known that both the β-adrenoceptor and rhodopsin were interacting with G-proteins in a stimulus-dependent fashion, but the fact that both receptors also shared structural similarity was not anticipated. Since the sequencing of rhodopsin, the 7TM architecture was thought to be the hallmark of light-sensitive proteins, but now it became clear that this topology was likely to be a common structural feature of all GPCRs. This idea was confirmed in the following years by the cloning of an ever-increasing number of GPCRs. The GPCRs were cloned based on similarity with already cloned family members. Because of this required similarity with already cloned GPCRs, it took a while until any distantly related receptors were discovered.

Not all of the newly cloned receptors had a known function or a known natural ligand. These receptors for which the sequence was known, but the function and/or endogenous ligand was unknown, are termed "orphan receptors". The first example of an orphan receptor was the clone "G21", which was isolated from a genomic DNA library shortly after the cloning of the β2-adrenoceptor (Kobilka et al. 1987a). This receptor was "deorphanized" by Fargin et al. (1988). The strategy to "deorphanize" the orphan receptors was to express the orphan GPCR of interest in eukaryotic cells by DNA transfection, and use the membranes of these cells as targets for testing the binding abilities of potential ligands. This strategy is now known as reverse pharmacology (Libert et al. 1991; Mills and Duggan 1994). Vassart et al. were the first to use PCR for finding new GPCRs (Libert et al. 1989), and continued to work on deorphanizing GPCRs till today, with great success (Brézillon et al. 2001, 2003; Parmentier and Detheux 2006).

With the availability of a large body of cloned receptors, the focus of the GPCR research field shifted toward unraveling the structural features responsible for the many aspects of receptor function. By using a wide range of molecular genetics techniques, ranging from site-directed mutagenesis approaches to the creation of chimeric receptors, the regions of the receptors responsible for G-protein coupling and ligand-binding were determined (Ostrowski et al. 1992; Strader et al. 1994). The coupling of G-proteins was attributed to the intracellular loops, whereas the binding of ligands, depending on the receptor subtype, was found to take place in the outward regions of the membrane-spanning domain and sometimes also partly in the extracellular domains.

The mutagenesis work provided a clarification of a previously unexplained phenomenon whereby many GPCRs were known to possess an intrinsic high background activity in the absence of a ligand, and led to the development of an interesting new concept: constitutive activity (Costa et al. 1992). Cotecchia et al. (1992) had created a chimeric receptor by replacing four residues of the third intracellular loop of the β2-adrenoceptor with residues from the α1b-adrenoceptor. This chimeric receptor had a surprising feature; it had the ability to signal measurably in the absence of an agonist. This feature was termed constitutive activity, and it was defined as ligand-independent activity resulting in the production of a second messenger even in the absence of an agonist. It was found that virtually any substitution in that region led to increased constitutive activity (Kjelsberg et al. 1992). It was hypothesized that this activity arose due to the fact that these mutations disrupted interactions that normally keep the receptor in an inactive state (Lefkowitz et al. 1993). It was found that many diseases were linked to naturally occurring mutations in GPCRs that resulted in constitutively active receptors (Seifert and Wenzel-Seifert 2002).

Constitutive activity is not limited to mutant receptors. In fact, even before Cotecchia described the first constitutively active mutant, Costa and Herz (1989) had already described the constitutive activity of the wild-type δ-opioid receptor. Since then, numerous observations have indicated that the basal activity of a wild-type GPCR might vary from totally inactive to fully active, depending on the nature of the GPCR.

Inverse agonism was already long known, but the discovery of (mutant inducible) constitutive activity made this concept accessible to experimentation in the GPCR field. Inverse agonists are a class of ligands that are capable of reversing the constitutive activity by stabilizing the inactive state of the receptor (Costa and Herz 1989; Bond et al. 1995). Before the discovery of constitutive activity in GPCRs, these compounds had been indistinguishable from antagonists (Kenakin 2001, 2002).

By the mid-1980s, it had become clear that both rhodopsin and the β2-adrenoceptor were phosphorylated in a stimulus-dependent way (Wilden and Kühn 1982; Stadel et al. 1983). The phosphorylation seemed to be related to the process of receptor inactivation or desensitization. A small family of proteins, the G-protein-coupled receptor kinases (GRKs), was found to perform this phosphorylation (Pitcher et al. 1998).

In 1986, Wilden and Kuhn reported a small protein that bound to phosphorylated rhodopsin, leading to steric exclusion of transducin (Wilden et al. 1986). This protein was named arrestin.

In a search for similar mechanisms for the β2-adrenoceptor, Lefkowitz et al. (Benovic et al. 1987) found that the GRKs alone were not sufficient for desensitization, and that another factor was needed. They found this factor and named it β-arrestin (Lohse et al. 1990), after the arrestin compound found by Wilden and Kühn. Phosphorylation of the receptor stimulates the binding of β-arrestin, leading to steric exclusion of G-proteins, inhibiting further signaling. These two families, the GRKs and the arrestins, appear to regulate essentially all of the seven TM receptors. These proteins share, together with the heterotrimeric G-proteins, the ability to interact virtually universally with all of the receptors in a stimulus-dependent fashion.

For a decade, the structure of bacteriorhodopsin was the only structure available that remotely resembled anything like a GPCR. In 1990 low-resolution electron cryo-microscopic models of bacteriorhodopsin were published (Henderson et al. 1990). Hibert et al. (1991) and Dahl et al. (1991) were the first to produce three-dimensional (3D) models for GPCRs. These models were built using bacteriorhodopsin as a template. When the bovine rhodopsin structure became available (Palczewski et al. 2000), it was seen that these models were very imprecise (Oliveira et al. 2004), but for many years they were the best bioinformatics could do, and they certainly helped the entire GPCR field think about sequence/structure–function relations (Oliveira et al. 1994). Many mutation studies were guided by these first (poor) models, and they aided the studies aimed at elucidating the function per residue.

August 4, 2000 is an historical date in the GPCR field. On that day, the structure of bovine rhodopsin (Palczewski et al. 2000) became available, providing researchers with the first high-resolution crystal structure of a GPCR. The X-ray structures of bovine rhodopsin and bacteriorhodopsin are significantly different, and the structure of bovine rhodopsin provided some interesting and some unexpected features

(Oliveira et al. 2004). The structure showed an 8th helix that immediately after helix VII sticks out parallel to the cytosolic membrane surface, exactly as was predicted by Oliveira et al. (1999) a year earlier. The TM helices were quite irregularly shaped when compared to those of bacteriorhodopsin; helix II contained an α-bulge, and a part of helix VII was shaped as a 3_{10} helix. Moreover, sequence alignments and modeling studies reveal that it is likely that the α-bulge in helix II is not always in the same place, or even present at all (Bywater 2005). In contrast to popular belief, the loops did not all stick-out into the solvent, as the extracellular loop IV–V formed two β-hairpins that were folded on top of the retinal, between the helices. The first GPCR structure has not yet had the impact on the GPCR field that we, before August 2000, expected that it would have. This is probably caused by the fact that most models, despite their sometimes great imprecision, provided enough structural information to explain most experiments, and because we simply do not yet know enough about GPCRs to fully appreciate all the things that this structure taught us.

The end of 2007 was the start of a small explosion of structural information on GPCRs. With the publication of four structures of β-adrenoceptors (Cherezov et al. 2007; Rasmussen et al. 2007; Hanson et al. 2008; Warne et al. 2008) and one of the human adenosine receptor (Jaakola et al. 2008), nearly a decade of structural silence had come to a close. At a first glance, all these structures look highly similar with eight well-superposable helices. The largest differences were observed in loop IV–V. As the loop IV–V is intensively involved in crystal packing contacts in all known structures, it is not known yet whether these differences are real, and whether the differences have a functional importance.

GPCR signaling was long thought to consist of one ligand, activating one monomeric GPCR, affecting downstream adaptor proteins via one heterotrimeric G-protein. However, over the years, the idea that GPCRs function as oligomers got increasing amounts of experimental backing. The first suggestions that GPCRs might form oligomers dates from the beginning of the 1980s, when Conn et al. (1982a,b) used a bivalent antibody that had a gonadotrophin-releasing hormone (GnRH) antagonists attached. Their data implied that this "conjugated" antagonist might be capable of acting as an agonist that promoted "microaggregates" of GnRH receptor (GnRHR) leading to biphasic regulation of receptor surface expression (Conn et al. 1982a,b). Evidence from ligand binding and studies of inhibition using free TM helices provide strong support for the notion that GPCRs act as dimers. A peptide derived from a $\beta2$-adrenoceptor TM domain inhibits both receptor dimerization and activation (Hebert et al. 1996) or possibly even oligomers (Hébert and Bouvier 1998). It would take a long time before the first direct physical evidence for GPCR dimerization would become available. In 2003 Fotiadis showed the existence of rows of rhodopsin receptors within retinal disc membranes by using atomic force microscopy (Fotiadis et al. 2003). The dimer model has been furnished with support from bioinformatics studies (Dean et al. 2001) in which the interfaces between putative dimer interaction sites can be identified. These theoretical studies were anticipated by an early, seminal, experimental observation from the Lefkowitz laboratory (Limbird et al. 1975) of negative cooperativity in GPCRs. Lefkowitz probably anticipated the later finding of dimers as he concludes this article with: "Further investigation is required to elucidate the molecular mechanisms responsible for negatively cooperative site–site interactions and the possible regulatory functions they serve"; besides that we can now be fairly certain that dimer formation is at the basis of this phenomenon, this statement is still valid, 35 years later.

The simplest model for GPCR activation is shown in Fig. 1. Actually this model is an oversimplification, as the existence of dimers is not taken into account.

It should be kept in mind that all states interconvert very rapidly; the law of mass action ensures that binding of most molecules such as ligands, G-proteins, antibodies, arrestins, etc., makes the equilibrium shift in the direction of that bound state. This is nicely illustrated by experiments in which receptors are shown to bind their ligand more tightly if more G-protein is present (summarized in: Hamm 1998). The existence of allosteric modulators therefore is no surprise, and indeed these were found as early as in 1987 (Howard et al. 1987). The thermodynamic treatment of these modulators can still be improved greatly in the GPCR field. A large number of so-called allosteric modulators have been reported in the literature, and

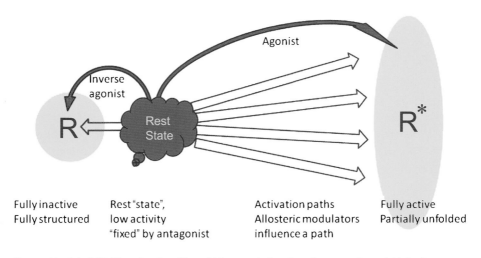

Fully inactive	Rest "state",	Activation paths	Fully active
Fully structured	low activity	Allosteric modulators	Partially unfolded
	"fixed" by antagonist	influence a path	

Fig. 1. **Model of GPCR activation.** This model does not take dimer formation or cytosolic protein binding into account, yet. Note that the rest state is only a "virtual" state that consists of a mixture of R and R*.

258

Christopoulos (e.g., Conn et al. 2009) has described them in a series of reviews. It is difficult – mainly because allosteric modulation was known in other receptor fields already before it was also discovered for GPCRs – to pinpoint the original discovery of this phenomenon.

3 GPCR structures

GPCRs are notoriously hard to crystalize. Proteins that are naturally present in an aqueous phase can usually be crystalized reasonably well, but the crystalization of membrane proteins has proven to be very difficult. Structure information is available for many GPCRs and the GPCRDB provides a list of about a hundred GPCR-related wwPDB entries. Most of these structure files, however, correspond to extramembrane domains and of those nearly all are extracellular. In addition to the inherent instability of GPCRs, one of the major problems is expressing GPCRs in large enough quantities in a form suitable for crystalization. For the determination of the structures that came available since 2007, it was in all cases necessary to jump through quite a large number of hoops to artificially stabilize the proteins.

In contrast to other fields where structures are leading experiments, in the GPCR field experimental data such as site-directed mutagenesis experiments, affinity labeling techniques, and ligand-binding studies (including so-called 2D mutations; van Galen et al. 1994; Horn et al. 1998, whereby the receptor and ligand are mutated in tandem) are leading the way in understanding the relation between GPCR sequence, structure, and function.

3.1 Rhodopsin

Rhodopsin was the first, and for a long time, the only GPCR structure available. In 2000 the first structure was published (Palczewski et al. 2000) and later a number of additional structures followed. The significance of these events can hardly be underestimated, as they provided a framework for the understanding of a large body of experimental data that was available for a long time already.

The fact that rhodopsin was the only structure available for 7 years is mainly due to the fact that rhodopsin is available in large amounts from bovine eye rod preparations, and due to the fact that rhodopsin is relatively stable compared to other GPCRs. Rhodopsin is not a ligand-mediated GPCR, and its sequence similarity to other class A receptors is fairly low. Despite having an enormous impact on the GPCR field, the rhodopsin structures were less suitable for modeling ligand-mediated structures than was initially anticipated (Oliveira et al. 2004).

259

The rhodopsin structure revealed a series of – sometimes very surprising – facts:

- Several helices are highly irregular, displaying highly uncommon features such as α-bulges and 3_{10} helix parts in the middle of a helix.
- An eighth helix, immediately after helix VII, indeed runs parallel to the membrane surface, as predicted by Oliveira et al. (1999).
- Although a few predictions to the contrary were made (Pardo et al. 1992), it was generally expected that the arrangement of GPCR helices would globally be similar to that of bacteriorhodopsin, and many models were constructed based on this concept (Dahl et al. 1991; Hibert et al. 1991; Donnelly and Findlay 1994). Even though the global helix packing agreed with in silico predictions, all homology models were too far off to have been of any use for structure driven drug design (Oliveira et al. 2004).
- By far the largest surprise in the bovine rhodopsin structure was the fact that the IV–V loop did not stick-out into the extracellular space, but rather was tucked away between the TM helices as a β-hairpin. This was unexpected, especially because no other protein in the wwPDB database contains a loop that folds inside/in-between the rest of the protein. This observation, however, did solve

Fig. 2. **Bovine rhodopsin structure (Palczewski et al. 2000, PDB ID 1F88).** Helices are shown in blue, strands in red, loops and very irregular helices in various shades of green, and retinal in yellow. Heavy metal atoms and sugar groups are removed for clarity.

one of the larger problems modelers had to deal with, namely, how to model the bridge between the cysteine in helix III that resides two helical turns deep into the membrane and the cysteine in the loop IV–V that was generally expected to be located outside the membrane region (Fig. 2).

3.2 Ligand-mediated GPCRs

It was not until the end of 2007 that the structures of the first ligand-mediated GPCRs were successfully determined. Almost simultaneously, two structures of the β2-adrenoceptor were published (Cherezov et al. 2007; Rasmussen et al. 2007), soon to be followed by the structures of the β1-adrenoceptor (Warne et al. 2008) and the α2-adenosine receptor (Jaakola et al. 2008). In contrast to the crystalization of rhodopsin, obtaining ligand-mediated GPCR crystals is very difficult, because GPCRs contain unstructured regions, and tend to cycle between various conformations spontaneously (Kobilka and Deupi 2007). To obtain high-quality crystals, the receptors must be very stable and of a conformational homogeneity. To achieve this, a number of different tricks were applied. This includes the use of antibody complexes (Rasmussen et al. 2007), fusion proteins (Cherezov et al. 2007; Jaakola et al. 2008), tight-binding ligands (Cherezov et al. 2007; Rasmussen et al. 2007; Jaakola et al. 2008; Scheerer et al. 2008; Warne et al. 2008) and stabilizing mutants (Warne et al. 2008). Although impressive, when interpreting a receptor structure model it is important to keep in mind the fact that in GPCRs there is an intricate relation between stability and the different aspects of receptor function.

When comparing the now available structural information, the first thing that one notices is the similar overall architecture of the TM segments. While the helices of the different structures are remarkably similar, the differences at the extracellular side are very large, even when the same structure is solved in two different labs. The differences are present both in the structure of this region as well as the interactions with the ligands.

It is important to keep in mind that when one looks at a crystal structure, this structure does not need to be in the orientation that the molecule would have under physiologic conditions. When looking at TM1 in the β1-adrenoceptor (Fig. 3), we can see a very strange helix conformation. This peculiarly shaped helix has not led to much excitement in the GPCR field, because everyone "knows" that such helices do not occur in vivo and that this strange conformation is caused by contacts in the crystal.

Crystal contacts do not always have such large effects, but it is wise to be aware of the effects these contacts can have and keep this effect in mind when interpreting a structure. We have shown the crystal packing interactions of a single receptor unit in the 2VT4 structure in Fig. 4. It can be seen that the amount of crystal contacts is very large.

Fig. 3.   Abnormal helix bending of TM1 (orange) in 2VT4.

When the crystal structures of the non-rhodopsin GPCRs came out, the helix in loop IV–V caused a lot of excitement, as this was a surprising feature. This region, however, is a region that has many crystal packing interactions. In Fig. 5, we have colored the regions that form contacts in the crystal for a number of GPCR structures. This figure clearly shows that the extramembrane areas, and especially the extracellular areas, are much involved in crystal packing interactions. Because of this, we must be cautious when interpreting these structural features, and be aware that they do not have to be an accurate representation of the "real" orientation of these parts. This implies that we cannot be certain about the spatial location of these areas, and, perhaps even worse, we cannot even be certain if they have any structure or that they are disordered and got ordered by binding either the ligand or the crystal partners, or both. On the other hand, the fact that the helix in loop IV–V is present in both the β1-adrenoceptor structure (2VT4) and the β2-adrenoceptor structure (2RH1) does provide support to the argument that these structural features in loop IV–V are real. However, loop IV–V is thought to have a number of different functional roles (ligand binding, ligand selectivity, and roles in activation), implying that loop IV–V is not present in one static conformation. It is more likely that this loop can exist in a number of conformations. This in turn implies that little energy is needed to change

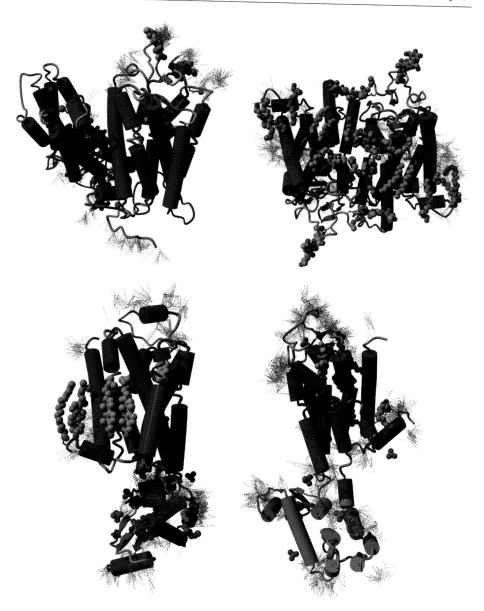

Fig. 4. **Top: 2 structures of bovine rhodopsin (left Palcewski 1F88, right Schertler 1GZM) Bottom left β2-adrenoceptor, bottom right adenosine α2a.** Helices are blue, strands red, loops and turns several shades of green, the ligand yellow, and purple bars indicate a contact with a partner in the crystal.

Fig. 5. The β1-adrenoceptor (2VT4) crystalizes as a tetramer. This figure uses the same coloring scheme as Fig. 4, but now the three partners in the crystalized asymmetric unit are shown as a purple cartoon and a 10 Å thick layer of residues in the crystal partners is drawn in thin purple lines.

the orientation, which brings us back to the argument that especially in these regions crystal contacts can have great influence on the orientation of these elements in the crystal.

The ligands bound to the adrenoceptor structures are both inverse agonists, meaning that they suppress the constitutive activity displayed by the receptor by stabilizing the inactive state. The ligand bound to the α2-adenosine receptor is an antagonist, meaning that this compound has no influence on the constitutive activity of the receptor, but just prevents the receptor from being activated by endogenous ligands.

The different effects of these compounds must be visible in the structures, and one would expect the inverse agonists, which have to alter the receptor's conformation, to have substantially more interactions with the receptor's active site than the antagonist, which basically just needs to sit "in the way".

The different binding modes of these ligands are illustrated in Fig. 6. This clearly shows that the antagonist compound does not have a lot of interactions within the binding pocket (it is even partially located outside the binding pocket), whereas the inverse agonists are firmly located deep in the binding pocket and having intensive interactions with the receptor.

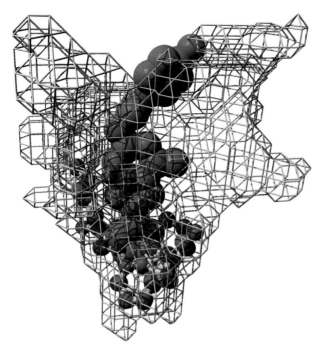

Fig. 6. **The active site cavity in the 2VT4 β1-adrenoceptor shown in yellow.** The inverse agonists in the two adrenoceptor structures (2RH1 and 2VT4), after superposing them using only the seven helices, are shown respectively as blue and red ball & stick models. The α2-adenosine antagonist in 3EML, again after superposing the structure on just the seven helices, is represented by purple balls. The inverse agonists have interactions deep down in the binding pocket while the antagonist seems to be "just in the way," higher up in the pocket.

In comparison with rhodopsin, the extracellular region of the adrenergic receptors is very open. The most prominent feature, which came as a surprise to many, is the existence of a short helical segment in the second extracellular loop. In the adenosine structure this small helix is not present. Rather, the presence of four disulfide bridges results in one part of the extracellular domain to be highly ordered, forming a small β-sheet, whereas another part of ECL2 was not visible in the electron density maps due to high flexibility. In both the adrenergic structures and the adenosine structure, the last few residues of ECL2 are located right above the ligand-binding pocket.

The extracellular region, especially the second extracellular loop, is thought to be of great importance for a lot of functional aspects. It contains a conserved cysteine bridge, which is the most conserved aspect throughout all GPCR families. This cysteine bridge, paradoxically, connects the most conserved region to the most variable part of the receptor (loop IV–V). Because of its conservation it must have an important function, which makes it all the more important to learn everything possible about its structure.

4 From sequence to structure

The sequences of the TM regions of GPCRs can be aligned very well, due to the presence of a number of very conserved residues, providing "anchors" for the sequence alignment. The loop regions, on the other hand, are notoriously difficult to align. Very little structural information is available for the loops, and the sequence variability between receptor subtypes is very large, both in terms of residue contents and in terms of size.

In the latest release of the GPCRDB (Horn et al. 2003), the stretch of residues around the conserved cysteine in loop IV–V (ECL2) has been included in the alignments. The previous release of the GPCRDB, dating back a few years, did not contain this information yet. In the hope to shed light on the strange conservation patterns of loop IV–V and the cysteine bridge, we have investigated the conservation and correlation patterns of the cysteines present in the second extracellular loop of a subfamily of the Class A receptors, the amine receptors.

4.1 The conserved cysteine bridge in the extracellular domain

It has been long observed that there are two conserved cysteine residues in the extracellular half of nearly all class A GPCRs. The number of cysteines present in the extracellular regions (including the highly conserved cysteine at the beginning of TMIII, C315) varies. It has been hypothesized that at least one covalent bond between two cysteines is conserved for all Class A GPCRs, and that many Class B and Class C also have a similar, conserved bridge. Looking at the alignment of the amine receptor family, both C315 and C470 are extremely well conserved.

For a number of GPCRs it has been experimentally determined that C315 and C470 form a disulfide bridge. Scholl and Wells (2000) showed that the effect of mutating either of the two cysteines in the adenosine A1 receptor leads to a complete loss of insertion into the membrane. The same experiment performed in the muscarinic ACM1 receptor in rat showed the same results (Savarese et al. 1992). Direct evidence for the existence of a C315–C470 disulfide bridge in the β2-adrenoceptor and the adenosine receptor came with the publication of their crystal structures.

4.2 Loop IV–V, cysteine bridges, and ligand binding

Cysteine 315 is located at the beginning of TMIII. Based on the common overall TM architecture of GPCRs with known structure we can be fairly certain where

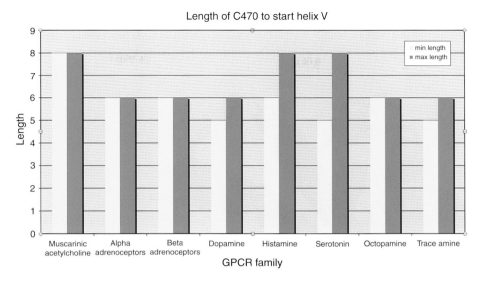

Fig. 7. **The length (in residues) of the stretch of residues from C470 to the start of helix V.** For each amine subfamily the minimum and maximum observed length is shown.

this residue is located with respect to other conserved elements. Because of the co-valent link between C315 and C470, we now can also position a part of the second extracellular loop in 3D space, despite the fact that this loop is extremely diverse in sequence. The fact that these two cysteines now can be anchored both in alignment space as well as in 3D space leads to new ways of thinking about the interactions of the binding pocket with agonists, antagonists, and inverse agonists.

We analyzed the length of the stretch of residues from C470 to the start of helix V for all proteins in the amine family. The length (in residues) is shown for each family within the amine receptor super family.

Figure 7 shows that the length of the last part of the loop IV–V is fairly short for all families within the amine receptor family. Variation in lengths within families is a result of differences in subfamilies, i.e., the shortest stretch observed within the histamine family has a length of six residues and is observed in the subfamily of the histamine type 2 receptors, whereas all other histamine subfamilies have an average length of eight residues.

The fact that the last part of loop IV–V is fairly short (5–8 residues) in all amine receptors, and the fact that the distance from the top of helix III (C315) to the start of helix V must be bridged by this stretch, implies that these residues are located on top of the ligand-binding pocket for all amine receptors, and this suggests that it must be involved in some aspects of ligand binding.

In all the GPCR structures available to date, the stretch of amino acids between C470 and the start of helix V interacts extensively with the ligand in the ligand-binding pocket. Due to the fact that both the position of C470 and the start of helix V are

Fig. 8. The stretch of amino acids from C470 to the start of helix V is shown in purple. The cysteine bridge between C470 and C315 is shown in yellow.

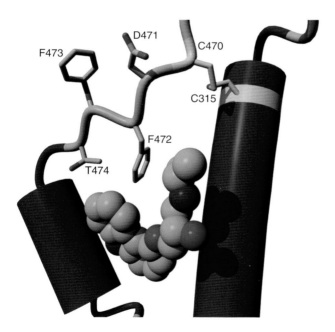

Fig. 9. Residues of the distal part of loop IV–V interacting with the bound ligand in the β2-adrenergic structure (2RH1).

known, we can now assign with great confidence a new part of the ligand-binding pocket. This is illustrated in Fig. 8, where the part of loop IV–V that is possibly interacting with the ligands is shown in purple.

All structural and much functional data that are available today show that residues 472 (sometimes also 473), and 474 interact with the bound ligand. Figure 9 shows the residues 472 and 474 in the β2-adrenoceptor structure 2RH1 (Cherezov et al. 2007) interacting with the ligand.

In the α2-adenosine receptor the situation is very similar. Kim et al. (1996) showed that the mutation E473A in the α2-adenosine receptor led to substantial changes in ligand-binding potential. This is in line with the α2-adrenoceptor structure, which shows that E473 interacts with the bound ligand (see Fig. 10). Note that here the orientation of the ECL2 is slightly different from that in the 2RH1 structure, exposing residues 472 and 473 to the ligand, whereas in the 2RH1 structure residues 472 and 474 are facing the ligand.

Shi and Javitch (2002) proposed that the loop IV–V plays an important role in ligand binding in the entire amine receptor family. Based on the cysteine bridge that is present in the rhodopsin crystal structure, they modeled the DRD2 receptor and found two residues that were likely to interact with the ligand. Experiments (mutations to cysteines, followed by sulfhydryl accessibility studies) performed on the dopamine-2 receptor by Shi and Javitch (2004) have shown that also in the DRD2 receptor the residues 472 and 474 are part of the ligand-binding pocket, confirming this hypothesis.

It appears that residues 471 and 473 might be involved in interactions with large ligands, or ligands that bind higher up in the receptor. Experiments of Wurch and

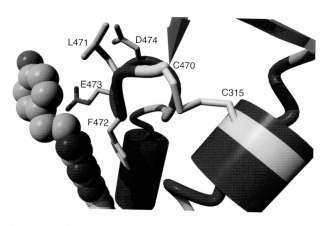

Fig. 10.   Residues of the distal part of loop IV–V interacting with the bound ligand in the α2-adenosine receptor (3EML).

Pauwels (2000) have shown that changing Q471 to leucine in canine 5HT1D enhanced ketanserin (large antagonist) affinity, while having no effect on 5HT (small agonist) binding. We think that this does not necessarily need to be an effect of the size of the ligand, but might also be related to the type of ligand: agonists enter deep down in the ligand pocket whereas antagonists – that only need to block entrance to the pocket to function perfectly – are expected to bind less deep in this pocket and thus are likely to contact other residues.

A number of studies have shown that mutations in the TM ligand-binding domain did have large effects on agonist binding characteristics, but that these mutations did not have any effect on antagonist binding characteristics (Townsend-Nicholson and Schofield 1994; van Galen et al. 1994). Zhao et al. (1996) showed that to convert the antagonist binding properties of the adenosine A1 receptor to those of the adenosine A2 receptor only three residues had to be mutated: residues 471, 472, and 473. This study also showed that mutations in the second extracellular loop do affect antagonist binding, but not agonist binding.

We now have seen examples of what is very likely to be a common theme in amine receptors. The conservation patterns for the various cysteine residues observed in the alignments accurately reflect the observed cysteine bridges in the available crystal structures and agree with available experimental data.

The residues after the conserved C470 are an integral part of the ligand-binding pocket and can influence ligand-binding properties as much as any ligand-binding residue in the TM domain. This knowledge is therefore of great significance for homology modeling and drug-docking studies. In addition, this feature offers new routes to elucidating the reasons for the previously unexplained specificity for certain GPCR subfamilies.

5 The future

Where do we go from here? To move into a more rational mode of drug design we should much better understand the sequence–structure–dynamics–function relations of GPCRs. What are the important questions that should be answered to get us really further in terms of this understanding? What is the role of GPCR dimers, is dimer formation a regulatory mechanism or is the main role increasing the combinatorics of signaling? Or both? How do sequence differences relate to functional differences? Why are receptors so promiscuous when it comes to binding G-proteins? The number of questions is still large, and a revival of the pharma industry critically depends on many answers in many of these fields.

The GPCR field has seen its share of bad models and models that were based on the selective use of references. Most models, either 3D models or mental models

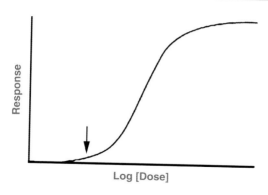

Fig. 11. **Generic dose–response curve.** Most biological and pharmacological experiments can, one way or another, be described by a plot like this one. The (inverse) agonist dose of zero is indicated by an arrow.

were backed up by carefully selected mutations. But a few things are clear. Figure 1 shows the generic GPCR situation. Most GPCRs show a low constitutive level of activity. Without any ligands they work at a fraction of their maximal activity. Figure 11 shows a very generic dose–response curve.

Inverse agonists move the arrow in Fig. 11 to the left and agonists to the right. All a pure antagonist needs to do is to avoid that agonists or inverse agonists can bind. So, a small molecule is most likely an antagonist if it binds in the upper half of the ligand-binding pocket where it only sits "in the way" but does not aid whatever processes (inverse) agonists trigger.

Much evidence points in the direction that GPCR activation requires partial disruption of the structure at the cytosolic side (Fanelli et al. 1999; Kim et al. 2004; Hornak et al. 2009). This disruption can be achieved by mutations, by over-expressing G-proteins, or by ligands, so that we can conclude that there is not one specific active conformation, but that there are many conformations that can bind and activate the G-protein; perhaps the active forms are partially unfolded, an hypothesis that is supported by the lack of visible structure in the cytosolic domains in all GPCR structures solved so far. The fact that the latter observation was predicted by Oliveira et al. (1999), before any GPCR structure information existed, shows the power of mental models for GPCR research.

Many models have been published over the years that place GPCR activation in schemas that are variations on the theme $R \leftrightarrow R^*$. These schemes can be rather complicated and sometimes involve multiple, interacting thermodynamic cycles. Whatever model we come up with, in the end amino acids will have to do the work, in other words, models should not disagree with realities such as the laws of thermodynamics. Unfortunately, in nearly all cases the authors of $R \leftrightarrow R^*$ models discuss

GPCR activation as a series of consecutive steps such as "ligand-binding causes a structural change that causes G-protein binding". Although useful for some lines of research, such models are wrong and often such a simplified model confuses more than that it clarifies. Obviously, ligand binding, G-protein binding, dimer formation, and even binding proteins involved in the down regulation all happen synchronously while some of these binding processes strengthen each other; e.g., positive cooperativity between agonist and G-protein binding because the agonist stabilizes the receptor conformation that is good for G-protein binding, and G-protein binding stabilizes the conformation that is good for agonist binding. In other cases, processes compete.

Another grand question is what actually happens residue by residue when a ligand or a G-protein binds? Given that GPCRs are constitutively active at a measurable fraction of their maximal activity (which should be read as that there is at any moment of time a certain fraction of all receptors fully active while the majority is inactive), it seems most likely that all processes that are involved in activation are on-going all the time. Ligand binding, G-protein binding, etc., just shifts those equilibriums. Do residues flip from one rotameric state to the other as suggested by Balasteros (Shi et al. 2002), or do whole helices move as was observed by Farrens et al. (1996), and in silico predicted by Fanelli et al. (1999) and later, in silico again coupled to ligand binding by Abagyan et al. (Katritch et al. 2009)? We suggest that there are a few mechanistic concepts shared between all GPCRs. One is the coupling to the G-proteins that, looking at the high level of promiscuity, must be done highly similarly by all GPCRs; perhaps via the Arg340 salt bridge with the conserved aspartic acid in helix V of the G-proteins, as suggested by Oliveira et al. (1999). The other common concept must relate to the ultra conserved cysteine bridge between the cysteine near the extracellular side of helix III and the cysteine in the loop IV–V. This cysteine bridge can "feel" if a G-protein binds via its partner in helix III, and it can "see" the ligand via its partner in the loop IV–V. We therefore suggest that, one way or another, this cysteine bridge plays an important role in the motions of amino acids, helices, and domains that are related with signaling.

We do not know yet why GPCRs form dimers. This can be related to regulation, to an extension of the signaling options, or, most likely, a combination of these two. Nevertheless, we can be certain that there must be "communication" between the monomers in the dimer. Both the combinatorics – and the regulation hypothesis require that the one monomer knows the state of the other monomer in terms of ligand binding and G-protein coupling. Given the fact that both ligand binding and G-protein binding can be "felt" by the conserved cysteine bridge, it does not seem very farfetched to hypothesize that this cysteine bridge is also involved in dimer communication. How this is done, and which residues are involved is still unknown.

We might speculate that the conserved Trp420 at the membrane surface of helix IV has a role in dimer formation. In that case, disturbance of the loop IV–V by whatever mechanism might also be "felt" in helix IV and then thus also in the dimer interface. However, a role for helix VIII in dimer communication also cannot be excluded. The location at opposite sides of the helix bundle of Trp420 and helix VIII makes it unlikely that both suggested mechanism operate in tandem.

There are enough questions to keep us all busy for decennia to come. Data from many different sources have played a role in much of the research we have cited in this article. We have collected most GPCR-related data in the GPCRDB. And we will keep working on this system. A thorough understanding of the sequence– struc-ture–dynamics–function relations of GPCRs is so important for our future quality of life that our whole effort would already pay off if just 1 day, one person is browsing the GPCRDB and gets one idea that brings us one step closer to answering one of the main questions left regarding these intriguing molecules.

References

Benovic JL, Kühn H, Weyand I, Codina J, Caron MG, Lefkowitz RJ (1987) Functional desensitiza-tion of the isolated beta-adrenergic receptor by the beta-adrenergic receptor kinase: potential role of an analog of the retinal protein arrestin (48-kDa protein). Proc Natl Acad Sci USA 84: 8879–8882

Bond RA and Ijzerman AP (2006) Recent developments in constitutive receptor activity and inverse agonism, and their potential for GPCR drug discovery. Trends Pharmacol Sci 27: 92–96

Bond RA, Leff P, Johnson TD, Milano CA, Rockman HA, McMinn TR, Apparsundaram S, Hyek MF, Kenakin TP, Allen LF (1995) Physiological effects of inverse agonists in transgenic mice with myo-cardial overexpression of the beta 2-adrenoceptor. Nature 374: 272–276

Brézillon S, Detheux M, Parmentier M, Hökfelt T, Hurd YL (2001) Distribution of an orphan G-pro-tein coupled receptor (JP05) mRNA in the human brain. Brain Res 921: 21–30

Brézillon S, Lannoy V, Franssen J, Le Poul E, Dupriez V, Lucchetti J, Detheux M, Parmentier M (2003) Identification of natural ligands for the orphan G-protein-coupled receptors GPR7 and GPR8. J Biol Chem 278: 776–783

Bywater RP (2005) Location and nature of the residues important for ligand recognition in G-protein coupled receptors. J Mol Recognit 18: 60–72

Caron MG, Srinivasan Y, Pitha J, Kociolek K, Lefkowitz RJ (1979) Affinity chromatography of the beta-adrenergic receptor. J Biol Chem 254: 2923–2927

Cerione RA, Strulovici B, Benovic JL, Strader CD, Caron MG, Lefkowitz RJ (1983) Reconstitution of beta-adrenergic receptors in lipid vesicles: affinity chromatography-purified receptors confer cate-cholamine responsiveness on a heterologous adenylate cyclase system. Proc Natl Acad Sci USA 80: 4899–4903

Cerione RA, Sibley DR, Codina J, Benovic JL, Winslow J, Neer EJ, Birnbaumer L, Caron MG, Lefkowitz RJ (1984) Reconstitution of a hormone-sensitive adenylate cyclase system. The pure beta-adren-ergic receptor and guanine nucleotide regulatory protein confer hormone responsiveness on the resolved catalytic unit. J Biol Chem 259: 9979–9982

Cherezov V, Rosenbaum DM, Hanson MA, Rasmussen SGF, Thian FS, Kobilka TS, Choi H, Kuhn P, Weis WI, Kobilka BK, Stevens RC (2007) High-resolution crystal structure of an engineered human beta2-adrenergic G-protein-coupled receptor. Science 318: 1258–1265

Conn PM, Rogers DC, McNeil R (1982a) Potency enhancement of a GnRH agonist: GnRH-receptor microaggregation stimulates gonadotropin release. Endocrinology 111: 335–337

Conn PM, Rogers DC, Stewart JM, Niedel J, Sheffield T (1982b) Conversion of a gonadotropin-releasing hormone antagonist to an agonist. Nature 296: 653–655

Conn PJ, Christopoulos A, Lindsley CW (2009) Allosteric modulators of GPCRs: a novel approach for the treatment of CNS disorders. Nat Rev Drug Discov 8: 41–54

Costa T and Herz A (1989) Antagonists with negative intrinsic activity at delta opioid receptors coupled to GTP-binding proteins. Proc Natl Acad Sci USA 86: 7321–7325

Costa T, Ogino Y, Munson PJ, Onaran HO, Rodbard D (1992) Drug efficacy at guanine nucleotide-binding regulatory protein-linked receptors: thermodynamic interpretation of negative antagonism and of receptor activity in the absence of ligand. Mol Pharmacol 41: 549–560

Cotecchia S, Ostrowski J, Kjelsberg MA, Caron MG, Lefkowitz RJ (1992) Discrete amino acid sequences of the alpha 1-adrenergic receptor determine the selectivity of coupling to phosphatidylinositol hydrolysis. J Biol Chem 267: 1633–1639

Dahl SG, Edvardsen O, Sylte I (1991) Molecular dynamics of dopamine at the D2 receptor. Proc Natl Acad Sci USA 88: 8111–8115

De Lean A, Stadel JM, Lefkowitz RJ (1980) A ternary complex model explains the agonist-specific binding properties of the adenylate cyclase-coupled beta-adrenergic receptor. J Biol Chem 255: 7108–7117

Dean MK, Higgs C, Smith RE, Bywater RP, Snell CR, Scott PD, Upton GJ, Howe TJ, Reynolds CA (2001) Dimerization of G-protein-coupled receptors. J Med Chem 44: 4595–4614

Dixon RA, Kobilka BK, Strader DJ, Benovic JL, Dohlman HG, Frielle T, Bolanowski MA, Bennett CD, Rands E, Diehl RE, Mumford RA, Slater EE, Sigal IS, Caron MG, Lefkowitz RJ, Strader CD (1986) Cloning of the gene and cDNA for mammalian beta-adrenergic receptor and homology with rhodopsin. Nature 321: 75–79

Donnelly D and Findlay JB (1994) Seven-helix receptors: structure and modelling. Curr Opin Struct Biol 4: 582–589

Fanelli F, Menziani C, Scheer A, Cotecchia S, De Benedetti PG (1999) Theoretical study of the electrostatically driven step of receptor-G-protein recognition. Proteins 37: 145–156

Fargin A, Raymond JR, Lohse MJ, Kobilka BK, Caron MG, Lefkowitz RJ (1988) The genomic clone G-21 which resembles a beta-adrenergic receptor sequence encodes the 5-HT1A receptor. Nature 335: 358–360

Farrens DL, Altenbach C, Yang K, Hubbell WL, Khorana HG (1996) Requirement of rigid-body motion of transmembrane helices for light activation of rhodopsin. Science 274: 768–770

Fotiadis D, Liang Y, Filipek S, Saperstein DA, Engel A, Palczewski K (2003) Atomic-force microscopy: rhodopsin dimers in native disc membranes. Nature 421: 127–128

van Galen PJ, van Bergen AH, Gallo-Rodriguez C, Melman N, Olah ME, Ijzerman AP, Stiles GL, Jacobson KA (1994) A binding site model and structure-activity relationships for the rat A3 adenosine receptor. Mol Pharmacol 45: 1101–1111

Gilman AG (1987) G-proteins: transducers of receptor-generated signals. Ann Rev Biochem 56: 615–649

Hamm HE (1998) The many faces of G-protein signaling. J Biol Chem 273: 669–672

Hanson MA, Cherezov V, Griffith MT, Roth CB, Jaakola V, Chien EYT, Velasquez J, Kuhn P, Stevens RC (2008) A specific cholesterol binding site is established by the 2.8 Å structure of the human beta2-adrenergic receptor. Structure 16: 897–905

Hanyaloglu AC and von Zastrow M (2008) Regulation of GPCRs by endocytic membrane trafficking and its potential implications. Ann Rev Pharmacol Toxicol 48: 537–568

Hargrave PA, McDowell JH, Curtis DR, Wang JK, Juszczak E, Fong SL, Rao JK, Argos P (1983) The structure of bovine rhodopsin. Biophys Struct Mech 9: 235–244

Hebert TE, Moffett S, Morello JP, Loisel TP, Bichet DG, Barret C, Bouvier M (1996) A peptide derived from a beta2-adrenergic receptor transmembrane domain inhibits both receptor dimerization and activation. J Biol Chem 271: 16384–16392

Henderson R and Unwin PN (1975) Three-dimensional model of purple membrane obtained by electron microscopy. Nature 257: 28–32

Henderson R, Baldwin JM, Ceska TA, Zemlin F, Beckmann E, Downing KH (1990) Model for the structure of bacteriorhodopsin based on high-resolution electron cryo-microscopy. J Mol Biol 213: 899–929

Hébert TE and Bouvier M (1998) Structural and functional aspects of G-protein-coupled receptor oligomerization. Biochem Cell Biol 76: 1–11

Hibert MF, Trumpp-Kallmeyer S, Bruinvels A, Hoflack J (1991) Three-dimensional models of neurotransmitter G-binding protein-coupled receptors. Mol Pharmacol 40: 8–15

Horn F, Bywater R, Krause G, Kuipers W, Oliveira L, Paiva AC, Sander C, Vriend G (1998) The interaction of class B G-protein-coupled receptors with their hormones. Receptors Channels 5: 305–314

Horn F, Bettler E, Oliveira L, Campagne F, Cohen FE, Vriend G (2003) GPCRDB information system for G-protein-coupled receptors. Nucleic Acids Res 31: 294–297

Hornak V, Ahuja S, Eilers M, Goncalves JA, Sheves M, Reeves PJ, Smith SO (2010) Light activation of rhodopsin: insights from molecular dynamics simulations guided by solid-state NMR distance restraints. J Mol Biol 396(3): 510–527

Howard MJ, Hughes RJ, Motulsky HJ, Mullen MD, Insel PA (1987) Interactions of amiloride with alpha- and beta-adrenergic receptors: amiloride reveals an allosteric site on alpha 2-adrenergic receptors. Mol Pharmacol 32: 53–58

Jaakola V, Griffith MT, Hanson MA, Cherezov V, Chien EYT, Lane JR, Ijzerman AP, Stevens RC (2008) The 2.6 Angstrom crystal structure of a human A2A adenosine receptor bound to an antagonist. Science 322: 1211–1217

Katritch V, Reynolds KA, Cherezov V, Hanson MA, Roth CB, Yeager M, Abagyan R (2009) Analysis of full and partial agonists binding to beta2-adrenergic receptor suggests a role of transmembrane helix V in agonist-specific conformational changes. J Mol Recognit 22: 307–318

Kenakin T (2001) Inverse, protean, and ligand-selective agonism: matters of receptor conformation. FASEB J 15: 598–611

Kenakin T (2002) Efficacy at G-protein-coupled receptors. Nat Rev Drug Discov 1: 103–110

Kent RS, De Lean A, Lefkowitz RJ (1980) A quantitative analysis of beta-adrenergic receptor interactions: resolution of high and low affinity states of the receptor by computer modeling of ligand binding data. Mol Pharmacol 17: 14–23

Kim J, Jiang Q, Glashofer M, Yehle S, Wess J, Jacobson KA (1996) Glutamate residues in the second extracellular loop of the human A2a adenosine receptor are required for ligand recognition. Mol Pharmacol 49: 683–691

Kim J, Altenbach C, Kono M, Oprian DD, Hubbell WL, Khorana HG (2004) Structural origins of constitutive activation in rhodopsin: role of the K296/E113 salt bridge. Proc Natl Acad Sci USA 101: 12508–12513

Kjelsberg MA, Cotecchia S, Ostrowski J, Caron MG, Lefkowitz RJ (1992) Constitutive activation of the alpha 1B-adrenergic receptor by all amino acid substitutions at a single site. Evidence for a region which constrains receptor activation. J Biol Chem 267: 1430–1433

Kobilka BK and Deupi X (2007) Conformational complexity of G-protein-coupled receptors. Trends Pharmacol Sci 28: 397–406

Kobilka BK, Frielle T, Collins S, Yang-Feng T, Kobilka TS, Francke U, Lefkowitz RJ, Caron MG (1987a) An intronless gene encoding a potential member of the family of receptors coupled to guanine nucleotide regulatory proteins. Nature 329: 75–79

Kobilka BK, Frielle T, Dohlman HG, Bolanowski MA, Dixon RA, Keller P, Caron MG, Lefkowitz RJ (1987b) Delineation of the intronless nature of the genes for the human and hamster beta 2-adrenergic receptor and their putative promoter regions. J Biol Chem 262: 7321–7327

Kobilka BK, Matsui H, Kobilka TS, Yang-Feng TL, Francke U, Caron MG, Lefkowitz RJ, Regan JW (1987c) Cloning, sequencing, and expression of the gene coding for the human platelet alpha 2-adrenergic receptor. Science 238: 650–656

Kovacs JJ, Hara MR, Davenport CL, Kim J, Lefkowitz RJ (2009) Arrestin development: emerging roles for beta-arrestins in developmental signaling pathways. Dev Cell 17: 443–458

Langley JN (1905) On the reaction of cells and of nerve-endings to certain poisons, chiefly as regards the reaction of striated muscle to nicotine and to curari. J Physiol 33: 374–413

Lefkowitz RJ, Cotecchia S, Samama P, Costa T (1993) Constitutive activity of receptors coupled to guanine nucleotide regulatory proteins. Trends Pharmacol Sci 14: 303–307

Libert F, Parmentier M, Lefort A, Dinsart C, Van Sande J, Maenhaut C, Simons MJ, Dumont JE, Vassart G (1989) Selective amplification and cloning of four new members of the G-protein-coupled receptor family. Science 244: 569–572

Libert F, Vassart G, Parmentier M (1991) Current developments in G-protein-coupled receptors. Curr Opin Cell Biol 3: 218–223

Limbird LE, Meyts PD, Lefkowitz RJ (1975) Beta-adrenergic receptors: evidence for negative cooperativity. Biochem Biophys Res Commun 64: 1160–1168

Lohse MJ, Benovic JL, Codina J, Caron MG, Lefkowitz RJ (1990) beta-Arrestin: a protein that regulates beta-adrenergic receptor function. Science 248: 1547–1550

Maehle A (2004) "Receptive substances": John Newport Langley (1852–1925) and his path to a receptor theory of drug action. Med Hist 48: 153–174

May DC, Ross EM, Gilman AG, Smigel MD (1985) Reconstitution of catecholamine-stimulated adenylate cyclase activity using three purified proteins. J Biol Chem 260: 15829–15833

Mickey J, Tate R, Lefkowitz RJ (1975) Subsensitivity of adenylate cyclase and decreased beta-adrenergic receptor binding after chronic exposure to (minus)-isoproterenol in vitro. J Biol Chem 250: 5727–5729

Mills A and Duggan MJ (1994) Orphan seven transmembrane domain receptors: reversing pharmacology. Trends Biotechnol 12: 47–49

Mukherjee C, Caron MG, Coverstone M, Lefkowitz RJ (1975) Identification of adenylate cyclase-coupled beta-adrenergic receptors in frog erythrocytes with (minus)-[3-H] alprenolol. J Biol Chem 250: 4869–4876

Oliveira L, Paiva AC, Sander C, Vriend G (1994) A common step for signal transduction in G-protein-coupled receptors. Trends Pharmacol Sci 15: 170–172

Oliveira L, Paiva AC, Vriend G (1999) A low resolution model for the interaction of G-proteins with G-protein-coupled receptors. Protein Eng 12: 1087–1095

Oliveira L, Hulsen T, Lutje Hulsik D, Paiva ACM, Vriend G (2004) Heavier-than-air flying machines are impossible. FEBS Lett 564: 269–273

Ostrowski J, Kjelsberg MA, Caron MG, Lefkowitz RJ (1992) Mutagenesis of the beta 2-adrenergic receptor: how structure elucidates function. Ann Rev Pharmacol Toxicol 32: 167–183

Ovchinnikov YuA (1982) Rhodopsin and bacteriorhodopsin: structure–function relationships. FEBS Lett 148: 179–191

Overington JP, Al-Lazikani B, Hopkins AL (2006) How many drug targets are there? Nat Rev Drug Discov 5: 993–996

Palczewski K, Kumasaka T, Hori T, Behnke CA, Motoshima H, Fox BA, Le Trong I, Teller DC, Okada T, Stenkamp RE, Yamamoto M, Miyano M (2000) Crystal structure of rhodopsin: a G-protein-coupled receptor. Science 289: 739–745

Pardo L, Ballesteros JA, Osman R, Weinstein H (1992) On the use of the transmembrane domain of bacteriorhodopsin as a template for modeling the three-dimensional structure of guanine nucleotide-binding regulatory protein-coupled receptors. Proc Natl Acad Sci USA 89: 4009–4012

Parmentier M and Detheux M (2006) Deorphanization of G-protein-coupled receptors. Ernst Schering Found Symp Proc 2: 163–186

Pepitoni S, Wood IC, Buckley NJ (1997) Structure of the m1 muscarinic acetylcholine receptor gene and its promoter. J Biol Chem 272: 17112–17117

Pert CB and Snyder SH (1973) Opiate receptor: demonstration in nervous tissue. Science 179: 1011–1014

Pitcher JA, Freedman NJ, Lefkowitz RJ (1998) G-protein-coupled receptor kinases. Ann Rev Biochem 67: 653–692

Rall TW and Sutherland EW (1958) Formation of a cyclic adenine ribonucleotide by tissue particles. J Biol Chem 232: 1065–1076

Rasmussen SGF, Choi H, Rosenbaum DM, Kobilka TS, Thian FS, Edwards PC, Burghammer M, Ratnala VRP, Sanishvili R, Fischetti RF, Schertler GFX, Weis WI, Kobilka BK (2007) Crystal structure of the human beta2 adrenergic G-protein-coupled receptor. Nature 450: 383–387

Ritter SL and Hall RA (2009) Fine-tuning of GPCR activity by receptor-interactinG-proteins. Nat Rev Mol Cell Biol 10: 819–830

Rodbell M, Birnbaumer L, Pohl SL, Krans HM (1971) The glucagon-sensitive adenyl cyclase system in plasma membranes of rat liver. V. An obligatory role of guanylnucleotides in glucagon action. J Biol Chem 246: 1877–1882

Rognan D (2006) Ligand design for G-protein-coupled receptors. Wiley-VCH

Ross EM and Gilman AG (1977) Resolution of some components of adenylate cyclase necessary for catalytic activity. J Biol Chem 252: 6966–6969

Savarese TM, Wang CD, Fraser CM (1992) Site-directed mutagenesis of the rat m1 muscarinic acetylcholine receptor. Role of conserved cysteines in receptor function. J Biol Chem 267: 11439–11448

Scheerer P, Park JH, Hildebrand PW, Kim YJ, Krauss N, Choe H, Hofmann KP, Ernst OP (2008) Crystal structure of opsin in its G-protein-interacting conformation. Nature 455: 497–502

Scholl DJ and Wells JN (2000) Serine and alanine mutagenesis of the nine native cysteine residues of the human A(1) adenosine receptor. Biochem Pharmacol 60: 1647–1654

Seifert R and Wenzel-Seifert K (2002) Constitutive activity of G-protein-coupled receptors: cause of disease and common property of wild-type receptors. Naunyn Schmiedeberg's Arch Pharmacol 366: 381–416

Shi L and Javitch JA (2002) The binding site of aminergic G-protein-coupled receptors: the transmembrane segments and second extracellular loop. Ann Rev Pharmacol Toxicol 42: 437–467

Shi L and Javitch JA (2004) The second extracellular loop of the dopamine D2 receptor lines the binding-site crevice. Proc Natl Acad Sci USA 101: 440–445

Shi L, Liapakis G, Xu R, Guarnieri F, Ballesteros JA, Javitch JA (2002) Beta2 adrenergic receptor activation. Modulation of the proline kink in transmembrane 6 by a rotamer toggle switch. J Biol Chem 277: 40989–40996

Stadel JM, Nambi P, Shorr RG, Sawyer DF, Caron MG, Lefkowitz RJ (1983) Catecholamine-induced desensitization of turkey erythrocyte adenylate cyclase is associated with phosphorylation of the beta-adrenergic receptor. Proc Natl Acad Sci USA 80: 3173–3177

Strader CD, Fong TM, Tota MR, Underwood D, Dixon RA (1994) Structure and function of G-protein-coupled receptors. Ann Rev Biochem 63: 101–132

Sunahara RK, Niznik HB, Weiner DM, Stormann TM, Brann MR, Kennedy JL, Gelernter JE, Rozmahel R, Yang YL, Israel Y (1990) Human dopamine D1 receptor encoded by an intronless gene on chromosome 5. Nature 347: 80–83

Sutherland EW and Rall TW (1958) Fractionation and characterization of a cyclic adenine ribonucleotide formed by tissue particles. J Biol Chem 232: 1077–1091

Tobin AB (2008) G-protein-coupled receptor phosphorylation: where, when and by whom. Br J Pharm 153 (Suppl 1): S167–S176

Townsend-Nicholson A and Schofield PR (1994) A threonine residue in the seventh transmembrane domain of the human A1 adenosine receptor mediates specific agonist binding. J Biol Chem 269: 2373–2376

Walsh DA, Perkins JP, Krebs EG (1968) An adenosine 3′,5′-monophosphate-dependant protein kinase from rabbit skeletal muscle. J Biol Chem 243: 3763–3765

Warne T, Serrano-Vega MJ, Baker JG, Moukhametzianov R, Edwards PC, Henderson R, Leslie AGW, Tate CG, Schertler GFX (2008) Structure of a beta1-adrenergic G-protein-coupled receptor. Nature 454: 486–491

Wilden U and Kühn H (1982) Light-dependent phosphorylation of rhodopsin: number of phosphorylation sites. Biochemistry 21: 3014–3022

Wilden U, Hall SW, Kühn H (1986) Phosphodiesterase activation by photoexcited rhodopsin is quenched when rhodopsin is phosphorylated and binds the intrinsic 48-kDa protein of rod outer segments. Proc Natl Acad Sci USA 83: 1174–1178

Wurch T and Pauwels PJ (2000) Coupling of canine serotonin 5-HT(1B) and 5-HT(1D) receptor subtypes to the formation of inositol phosphates by dual interactions with endogenous G(i/o) and recombinant G(alpha15) proteins. J Neurochem 75: 1180–1189

Yamamura HI and Snyder SH (1974) Muscarinic cholinergic binding in rat brain. Proc Natl Acad Sci USA 71: 1725–1729

Zhao MM, Hwa J, Perez DM (1996) Identification of critical extracellular loop residues involved in alpha 1-adrenergic receptor subtype-selective antagonist binding. Mol Pharmacol 50: 1118–1126

LIST OF CONTRIBUTORS

Sanne Abeln
Centre for Integrative Bioinformatics
VU University Amsterdam
1081 HV Amsterdam, The Netherlands

Pierre Baldi
School of Information and Computer
Sciences, Institute for Genomics and
Bioinformatics, University of California-
Irvine, Irvine, CA 92697-3435, USA
E-mail: pfbaldi@ics.uci.edu

Patrick Barth
Department of Pharmacology, Baylor
College of Medicine, One Baylor Plaza
Houston, TX 77030, USA
E-mail: patrickb@bcm.tmc.edu

Lisa Bartoli
Biocomputing Group, Bologna
Computational Biology Network
University of Bologna
40126 Bologna, Italy

Nir Ben-Tal
Department of Biochemistry and
Molecular Biology, George S. Wise
Faculty of Life Sciences, Tel Aviv
University, Ramat Aviv 69978, Israel

Robert P. Bywater
Medical Centre, CMBI, NCMLS,
Radboud University Nijmegen
6525 GA Nijmegen, The Netherlands

Rita Casadio
Biocomputing Group, Bologna
Computational Biology Network
University of Bologna
40126 Bologna, Italy
E-mail: casadio@biocomp.unibo.it

Piero Fariselli
Biocomputing Group, Bologna
Computational Biology Network
University of Bologna, 40126 Bologna, Italy

K. Anton Feenstra
Centre for Integrative Bioinformatics
VU University Amsterdam
1081 HV Amsterdam, The Netherlands

Dmitrij Frishman
Department of Genome Oriented
Bioinformatics, TU München
Wissenschaftszentrum Weihenstephan
85350 Freising, Germany
E-mail: d.frishman@wzw.tum.de

Angelika Fuchs
Department of Genome Oriented
Bioinformatics, TU München
Wissenschaftszentrum Weihenstephan
85350 Freising, Germany

Michael Y. Galperin
National Center for Biotechnology
Information, National Library of
Medicine, National Institutes of Health
Bethesda, MD 20894, USA

Erik Granseth
Department of Genome Oriented
Bioinformatics, TU München
Wissenschaftszentrum Weihenstephan
85350 Freising, Germany
E-mail: erikgr@gmail.com

Susan E. Harrington
Department of Biochemistry and
Molecular Biology, George S. Wise
Faculty of Life Sciences, Tel Aviv
University, Ramat Aviv 69978, Israel
E-mail: susan.e.harrington@gmail.com

Sikander Hayat
Center for Bioinformatics
Saarland University
66041 Saarbruecken, Germany

Volkhard Helms
Center for Bioinformatics
Saarland University
66041 Saarbruecken, Germany
E-mail: volkhard.helms@bioinformatik.
uni-saarland.de

Jaap Heringa
Centre for Integrative Bioinformatics
VU University Amsterdam
1081 HV Amsterdam, The Netherlands
E-mail: heringa@few.vu.nl

Jana R. Herrmann
Department für biowissenschaftliche
Grundlagen, TU München
Weihenstephaner Berg 3
85354 Freising and Munich Center for
Integrated Protein Science (CIPS^M)
Freising, Germany

Andreas Kirschner
Department of Genome Oriented
Bioinformatics, TU München
Wissenschaftszentrum Weihenstephan
85350 Freising, Germany

Dieter Langosch
Department für biowissenschaftliche
Grundlagen, TU München
Weihenstephaner Berg 3
85354 Freising and Munich Center for
Integrated Protein Science (CIPS^M)
Freising, Germany
E-mail: langosch@tum.de

Pier Luigi Martelli
Biocomputing Group, Bologna
Computational Biology Network
University of Bologna, 40126 Bologna
Italy

Madeleine Matias
Division of Biological Sciences
University of California at San Diego
La Jolla, CA 92093-0116, USA

Jennifer Metzger
Center for Bioinformatics
Saarland University
66041 Saarbruecken, Germany

Armen Y. Mulkidjanian
School of Physics
University of Osnabrück
49069 Osnabrück, Germany
E-mail: amulkid@uos.de

A.N. Belozersky Institute of Physico-
Chemical Biology, Moscow State
University, Moscow 119991, Russia

Laerte Oliveira
Medical Centre, CMBI, NCMLS
Radboud University Nijmegen
6525 GA Nijmegen, The Netherlands

Walter Pirovano
Centre for Integrative Bioinformatics
VU University Amsterdam
1081 HV Amsterdam, The Netherlands

Arlo Randall
School of Information and Computer
Sciences, Institute for Genomics and
Bioinformatics, University of California-
Irvine, Irvine, CA 92697-3435, USA

Milton H. Saier Jr
Division of Biological Sciences
University of California at San Diego
La Jolla, CA 92093-0116, USA

István Simon
Institute of Enzymology, BRC
Hungarian Academy of Sciences
1113 Budapest
Hungary

Eric I. Sun
Division of Biological Sciences
University of California at San Diego
La Jolla, CA 92093-0116, USA

Gábor E. Tusnády
Institute of Enzymology, BRC
Hungarian Academy of Sciences
1113 Budapest, Hungary

Stephanie Unterreitmeier
Department für biowissenschaftliche
Grundlagen, TU München
Weihenstephaner Berg 3
85354 Freising and Munich Center for
Integrated Protein Science (CIPS^M)
Freising, Germany

Gert Vriend
Medical Centre, CMBI, NCMLS
Radboud University Nijmegen
6525 GA Nijmegen, The Netherlands

Bas Vroling
Medical Centre, CMBI, NCMLS
Radboud University Nijmegen
6525 GA Nijmegen, The Netherlands

Bin Wang
Division of Biological Sciences
University of California at San Diego
La Jolla, CA 92093-0116, USA

Ming Ren Yen
Division of Biological Sciences
University of California at San Diego
La Jolla, CA 92093-0116, USA